Series: Advanced Sensors and Emerging Technologies for Smart Healthcare Services

Series Editor: Amit Kumar Tyagi

Smart Hospital Technologies

Advanced Sensors Driving Patient Care Innovations

Books in the Series

- Smart Hospital Technologies: Advanced Sensors Driving Patient Care Innovations
  Vivek Kumar Singh, Avinash Kumar Sharma and Tuan Anh Nguyen (eds.), 2027

Series: Advanced Sensors and Emerging Technologies for Smart Healthcare Services

Series Editor: Amit Kumar Tyagi

Smart Hospital Technologies

Advanced Sensors Driving Patient Care Innovations

Editors

Vivek Kumar Singh

Department of Computer Science & Engineering
Sharda School of Computing Science & Technology
Sharda University
Uttar Pradesh, India

Avinash Kumar Sharma

Department of Computer Science & Engineering
Sharda School of Computing Science & Technology
Sharda University
Uttar Pradesh, India

Tuan Anh Nguyen

Vietnam Academy of Science and Technology
Caugiay, Hanoi, Vietnam

CRC Press
Taylor & Francis Group
Boca Raton London New York

CRC Press is an imprint of the
Taylor & Francis Group, an **informa** business
A SCIENCE PUBLISHERS BOOK

Cover credit: Image created by a draftsman using Magnific, an AI platform, with a valid subscription and permission for use.

First edition published 2027
by CRC Press
2385 NW Executive Center Drive, Suite 320, Boca Raton FL 33431

and by CRC Press
4 Park Square, Milton Park, Abingdon, Oxon, OX14 4RN

Library of Congress Cataloging-in-Publication Data (applied for)

ISBN: 978-1-041-32658-8 (hbk)
ISBN: 978-1-041-32659-5 (pbk)
ISBN: 978-1-003-78385-5 (ebk)

DOI: 10.1201/9781003783855

Typeset in Times New Roman
by Prime Publishing Services

Preface to the Series

Today the convergence of advanced sensors and emerging technologies has increased in many applications and reshaped the healthcare industries, especially how healthcare systems function, diagnose, and deliver personalized care. This book is under the book series "*Advanced Sensors and Emerging Technologies for Smart Healthcare Services*," which provides the essence of such rapid transformation and shows the pathway toward a more intelligent, responsive, and patient-centered healthcare future. As we know, over the past decade, the growing convergence between biomedical engineering, data analytics, and information technology has made a major revolution in healthcare monitoring, diagnostics, and therapeutic innovation, etc. The objective of developing a book under this book series was to bring together diverse research perspectives and interdisciplinary insights that drive this rapidly evolving field.

As healthcare moves from hospital-centered models to continuous, connected ecosystem-based care, these technologies empower medical professionals to monitor patients in real time, predict disease progression, and ensure timely intervention while reducing costs and human errors. This book series seeks to bridge the gap between theoretical research and practical applications in smart healthcare solutions. It offers a platform for scholars, engineers, clinicians, and technologists to explore the rich potential of sensor technologies, artificial intelligence, Internet of Things (IoT), wearable systems, and machine learning in improving healthcare quality.

In recent years, wearable sensors, flexible electronics, and artificial intelligence (AI) have initiated a major shift toward predictive and preventive medicine. The ability to collect and analyze major physiological data continuously has redefined how diseases are detected and managed. But still, integrating these technologies into clinical practice faces unique challenges concerning data security, interoperability, and user acceptability (in general). This book in this book series focused on these challenges while offering design methodologies, frameworks, and models to guide innovation responsibly.

Each volume of this book series will focus on a specific theme, ranging from biosensors and nanotechnology to remote diagnostics, digital therapeutics, robotics, and augmented reality in healthcare. Together, this book provides a comprehensive understanding of both the enabling technologies and their societal, regulatory, and ethical dimensions.

The contributors in this book are renowned researchers who come from academia and industry and provide their deep insights into state-of-the-art developments, challenges in implementation, and future possibilities toward healthcare.

innovation and clinical practice, contributing to improved patient outcomes and more resilient healthcare systems.

The editors sincerely acknowledge the contributions of all authors, reviewers, and the publishing team for their efforts in bringing this volume to fruition. We hope this book will inspire further research, collaboration, and innovation in the evolving domain of smart and sensor-enabled healthcare.

Contents

Section 3 Applications and Use Cases

Section 4 Challenges and Opportunities

Section 1

Basic and Fundamentals

of holistic IoMT-based healthcare frameworks that unify technological innovation with a human-centric approach and integrate ethical governance with environmental sustainability creates a need to develop a paradigm that focuses on all these significant aspects. Significant past research has been carried out, fostering the systematic development of IoMT components and their device efficiency, which has rendered a need to develop a well-inclusive and comprehensive framework that not only addresses the technical aspects but also aligns the socio-technical paradigms of the diverse healthcare 5.0 ecosystem.

The present research work is carried out to address the critical void in the literature by developing a well-inclusive and multi-dimensional framework for IoMT-based healthcare systems, which integrates the aspects of cognisant interconnection, safety and confidentiality, user-centric engagement, and environmentally responsible healthcare systems. The purpose of this research is to move beyond technical aspects and establish a cohesive model that ensures equity, resilience, and long-term viability in next-generation healthcare 5.0 systems. The study synthesises the established foundational theories and proposes a framework that provides a structured approach for the development, design, and governance of IoMT-based healthcare structures. The significance of the study lies in redefining the aspect of IoMT processes from a mere data collection tool to enabling it as an integral component of an ethical, responsive, and adaptive healthcare system.

The expected outcome of the study involves the augmentation of AI-driven interoperability to harmonise healthcare data sources, develop a decentralised security architecture, and propose an empathic design that ensures the aspects of accessibility, inclusivity, and user engagement. The proposed framework introduces a novel concept of a circular health model that integrates the sustainability aspect into the IoMT ecosystem by addressing the issues of energy efficiency and carbon-neutral health infrastructure. Such an innovative approach is significant for ensuring health equity and enabling environmentally responsible solutions to extend healthcare quality to underserved and remote populations. The insight of the study enables technology developers, government systems, healthcare providers, and regulatory policymakers to deploy secure, patient-centred, and future-ready healthcare approaches. This study provides a strategic roadmap for embedding IoMT devices that supports regulatory compliance and data protection within national digital health policy systems. The present research work aims to contribute to the advancement of contemporary healthcare systems, which are facing challenges concerning cyber threats, health disparities, and sustainability issues. It aims to bridge the gap between technological potential and practical implications by evolving a novel intelligent health system in healthcare 5.0.

2. Review of Literature

The present study considers three realms of literature concerning (1) evolution of healthcare technologies and role of IoMT in Healthcare 5.0, (2) security, privacy, and decentralisation processes in Healthcare 5.0, (3) human-centric and sustainable IoMT systems for future healthcare. These can be elucidated as follows:

2.1 Evolution of Healthcare Technologies and the Role of IoMT in Healthcare 5.0

The role of Internet of Medical Things (IoMT) in healthcare 5.0 has been extensively explored and given prime importance in previous literature. The healthcare system around the globe has followed a long pathway wherein phase 1 was characterised by elementary medical practices, followed by maintenance of electronic health records in healthcare 2.0 systems (Bolhasani et al. 2021, Khan et al. 2023, Prabha et al. 2024). The introduction of connectivity systems and automated processes was followed in healthcare system 3.0, whereas advanced IoT and digitalisation processes were followed in healthcare 4.0 systems. The contemporary healthcare system, i.e., healthcare 5.0, intends to integrate and leverage IoMT processes, blockchain, artificial intelligence, and real-time data analytics in healthcare systems (Riaz et al. 2025, Ullah et al. 2025). Such sophisticated technologies are used to address numerous challenges in healthcare systems, specifically aiming for the detection of cancer cells, the identification of tumour zones, and the prediction of various minor and major diseases in human as well as animal bodies (Dutta and Puthal 2024, AlShorman et al. 2020). Certain studies emphasised the transformative role of the Internet of Medical Things by appreciating the advantages of healthcare smart devices. Such processes are making a significant shift towards Industry 5.0 by amalgamating healthcare systems with human–machine collaboration (Javed et al. 2025, Khan et al. 2023, Anwar and Prasad 2018). Such processes enable personalised care, real-time patient monitoring, and can lead to a proactive healthcare ecosystem with effective patient management. Few studies outlined an adequate framework for cancer patient care by exemplifying the efficiency of IoMT devices in communication as well as continuous monitoring between healthcare practitioners and cancer patients (Mohammed and Aljanabi 2023, Sharma et al. 2023). The framework was relevant in emphasising the adequacy of IoMT processes in transforming healthcare 5.0 by bridging the gaps between patients and hospitals, increasing care efficiency, and reducing the administrative tasks of the healthcare authorities.

The aspects of data privacy and resilience in IoMT-based healthcare have been extensively explored in past literature concerning federated learning and the use of blockchain systems (Almogadwy and Alqarafi 2025, Natarajan et al. 2023, Mbunge et al. 2021). A paradigm of federated learning was developed, which focused on developing decentralised patient monitoring and a secure system that not only mitigates the risk but also enables suitable data processing and storage (Rahman et al. 2025, Dutta and Puthal 2024, Mukhopadhyay et al. 2024). Additionally, certain studies focused on developing a sustainable healthcare system, which is based upon a synergistic paradigm between blockchain and IoMT processes. An intelligent healthcare decision-making system was proposed in established literature, which can efficiently work in heterogeneous environments (Sadeghi and Mahmoudi 2024, Sharma et al. 2023, Khan et al. 2023). In this regard, an integrated blockchain and fuzzy logic framework was outlined, which integrated cloud computing with AI and IoMT systems, aimed at developing an adequate healthcare system. Few studies provided strong evidence by focusing on the technical aspects of IoMT-based healthcare systems (Javed et al. 2025, Rahman et al. 2025, Mbunge et al. 2021).

The outcomes of such studies focused on developing system security and addressing the accuracy of IoMT-based models, while certain studies underscored developing a suitable user experience, paying due consideration to ethical aspects and improving the user experience (Ullah et al. 2025, AlShorman et al. 2020, Anwar and Prasad 2018). It is therefore concluded that a gap exists in developing a standardised framework that provides due guidance on the adequate deployment of IoMT-based healthcare technologies. The past reviewed literature, although it provided a robust foundation on IoMT-enabled healthcare 5.0, indicates that there is a need to develop a holistic, secure paradigm that aligns human expectations with modern healthcare delivery systems. It is identified that there is a need to develop a human-centred IoT system that not only engages and involves patient care, but also involves aspects of user-friendliness, ease of use, and accessibility.

2.2 Security, Privacy, and Decentralisation Processes in IoMT-based Healthcare 5.0

Adequate possibilities exist in the modern healthcare systems that enable remote diagnostics, personalised treatment plans, and real-time patient monitoring (Khubrani and Alam 2021, AlShorman et al. 2020). Such a scenario is possible by integrating the Internet of Medical Things into healthcare systems 5.0 (Riaz et al. 2025, Gupta et al. 2020, Wazid et al. 2019). However, the past literature established that enabling such systems will amplify concerns related to data, decentralisation, privacy, and security of IoMT-enabled devices (Mukhopadhyay et al. 2024, Samadhiya et al. 2024, Shukla et al. 2021). Various such challenges were highlighted, and it was underscored that there is a need to develop advanced blockchain-based cybersecurity systems that adequately mitigate various vulnerabilities in IoMT-based healthcare systems (Rashid and Nemati 2024, Asri et al. 2023, Khan et al. 2023). Few studies proposed an integrated federated learning paradigm that not only enables secure and decentralised patient monitoring but also focuses on enhancing data privacy and puts more reliance on centralised data repository systems (Rahman et al. 2025, Alkhalifa et al. 2024, Mukhopadhyay et al. 2024). The synergistic framework for Blockchain and Federated Learning must emphasise on preserving privacy at all levels. A decentralised architecture was proposed that improves trust among stakeholders and augments the data sovereignty of IoMT healthcare-based systems (Javed et al. 2025, Mbunge et al. 2021, Shukla et al. 2021).

Certain studies focused on developing smart contracts and immutable technology systems that secure IoMT data and prevent its tampering and unauthorised access (Rashid and Nemati 2024, Asri et al. 2023). The findings of such studies align with the broad consensus of AI-based systems that aim to develop a robust framework for ensuring auditability and data integrity in modern healthcare ecosystems (Singh 2025, Alam et al. 2024). A few divergent approaches were also witnessed in the established literature concerning the implementation of well-secured IoMT healthcare systems. Few studies focused on developing blockchain and AI e-health systems, few on developing interpretable machine learning models, whereas some studies underscored developing multi-dimensional IoMT security systems (Alkhalifa et al. 2024, Almalki et al. 2024, Gupta et al. 2020, Rashid and Nemati 2024,

Wazid et al. 2019). The ultimate objective of such contrasting frameworks is to devise a system that enables data encryption, access control, and threat detection on a simultaneous basis. There is a need to propose a comprehensive secure system that integrates intrusion detection, federated learning, and handles scalability issues. The established literature highlighted various gaps in the framework of IoMT healthcare systems. The gaps were specifically related to standardisation and compliance regarding regulatory frameworks (Alatawi 2023, Almogadwy and Alqarafi 2025, Rashid and Nemati 2024, Sadeghi and Mahmoudi 2024). It was emphasised that data integrity and privacy are integral in healthcare regulations, and there is a need to propose an authentication model based upon blockchain systems for IoMT healthcare (Khubrani and Alam 2021, Mukhopadhyay et al. 2024, Ullah et al. 2025, Wazid et al. 2019). Few studies outlined the ethical and legal implications of implementing the various healthcare regulations and found that there is a lack of standardised models that counter emerging threats, secure data analytics, and genetics in IoMT-based healthcare systems (Gupta et al. 2020, Mathkor et al. 2024, Rahman et al. 2025, Shukla et al. 2021, Singh 2025). The outcomes of the past studies focused on exploring various malware detection techniques in IoMT systems, but such exploration failed to accommodate the emerging threats in decentralised blockchain-integrated IoMT systems (Alkhalifa et al. 2024, Riaz et al. 2025, Samadhiya et al. 2024, Wazid et al. 2019). It is highlighted that there is a need to foster a context-specific security paradigm that accommodates unique vulnerabilities of IoMT devices concerning real-time data transmission, resource constraints, and device heterogeneity.

2.3 Human-centric and Sustainable IoMT Systems for Future Healthcare

An increasing emphasis has been placed in past literature concerning sustainable and human-centric healthcare systems, and, in this regard, patient needs shall be suitably aligned with technological innovation by considering the aspects of usability and long-term ecological integrity (Akram et al. 2025, Gomathi et al. 2023, Shukla et al. 2021). The past literature emphasised the significance of framing healthcare IoMT systems that are not only technologically advanced but also contain integral aspects of ethics, environmental responsibility, and user-friendliness (Alkhalifa et al. 2024, Bolhasani et al. 2021, Prabha et al. 2024). Certain studies underscore the development of a human-centred, IoMT-based health monitoring framework that considers the aspects of personalised care delivery and user engagement (Anwar and Prasad 2018, Bharadwaj et al. 2021, Bolhasani et al. 2021). A significant shift has been witnessed from device-centric models to human-centric models, where due accord was given to edge computing, AI systems, and wearable sensors (Alam et al. 2024, Javed et al. 2025, Natarajan et al. 2023). All such processes integrate the IoMT processes and are guided by health literacy and ethical considerations (Khubrani and Alam 2021). Certain studies recognised the convergence of IoMT systems and AI processes as a key enabler of adaptive healthcare monitoring systems that are intelligent and empathic for developing the future of healthcare systems (Almogadwy and Alqarafi 2025, Pradhan et al. 2021, Samadhiya et al. 2024).

service, and secure communication between patients and healthcare providers. This enabled the evaluation of the effectiveness of IoMT systems in providing healthcare outcomes related to patient engagement and hospital admissions (Almalki et al. 2024, Bolhasani et al. 2021). Subsequently, the complex adaptive system theory was considered, which is a dynamic approach that considers real-time decision-making during a health crisis. This theory has set an underpinning for enabling the comprehension of IoMT ecosystems involving various interconnected networks (Akram et al. 2025, Alkhalifa et al. 2024, Khubrani and Alam 2021). This theory supports the amalgamation of machine learning techniques, adaptive diagnostics, and automated interventions. Lastly, the current study considered the sustainability theory, which emphasises developing a socially equitable, economically viable, and environmentally responsible IoMT-based healthcare system, such as systems that support low-carbon healthcare infrastructure and promote inclusive health delivery models (Anwar and Prasad 2018, Bajpayi et al. 2024, Huang et al. 2024). The conceptual framework for the IoMT-based healthcare 5.0 systems, as showcased in Table 1, is collectively based on the said theories, which focus on developing a paradigm that is technology-advanced, user-centric, sustainable, and secure.

Table 1: Underlying theories governing IoMT-based healthcare 5.0 system.

Core Pillar of Healthcare 5.0	**Underlying Theories & Foundations**	**Key Objectives & Focus Areas**	**Technologies & Outcomes Enabled**
Technological Innovation	Diffusion of Innovation Theory; Complex Adaptive Systems Theory	Drive adoption of IoMT, ensure scalability, compatibility, and real-time adaptability in crises.	Wearables, real-time monitoring, ML-driven diagnostics, automated interventions, IoT scalability.
Decentralisation & Security	Information Systems Success Model; Socio-Technical Systems Theory	Ensure secure, ethical, high-quality data exchange; balance tech with human & ethical factors.	Blockchain-secured comms, patient-provider service quality, secure interoperability, trust systems.
Human-Centric & Participatory Design	Socio-Technical Systems Theory; Diffusion of Innovation (usability & literacy aspects)	Enhance patient engagement, usability, health literacy, and inclusive co-design with stakeholders.	Emotionally intelligent interfaces, participatory development, adaptive UX, reduced admission rates.
Sustainability & Equity	Sustainability Theory (Triple Bottom Line: Social, Economic, Environmental)	Build low-carbon, resource-efficient, equitable, and long-term viable healthcare ecosystems.	Energy-efficient IoMT, circular design, inclusive delivery models, carbon-neutral infrastructure.

Source: Author's creation

4. Proposed Framework for Holistic Health Intelligence in Healthcare 5.0 System

A transformative evolution is witnessed in the modern healthcare ecosystem, where it is found that conventional medical practices are transitioning to technologically enabled and data-driven systems. The emergence of the healthcare 5.0 system represents a paradigm shift, which amalgamates sophisticated technologies such as blockchain systems, artificial intelligence, real-time data analytics, and Internet of Medical Things. The ultimate purpose of this upsurge is to provide a secure, sustainable, and more personalised healthcare ecosystem. The past literature has underlined significant foundational aspects of IoMT and its integration in the healthcare systems. It was witnessed that the role of IoMT in healthcare is increasing due to remote diagnostics, rise in diseases, and adoption of decentralised data management systems. However, it is witnessed that there is a significant gap in developing a framework that is well-inclusive, comprehensive, and holistic. The proposed paradigm aligns human-centric values with technological innovation and, at the same time, considers the sustainability aspects. The framework proposed in the present study is a novel approach based on IoMT healthcare, transcending current implementations by including different aspects of participatory design, principles, adaptive intelligence, and secure data systems. The framework is aimed to empower the healthcare providers and ensure long-term viability in the complex, yet interconnected healthcare ecosystem.

The framework, as outlined in Figure 2, is based upon four significant aspects: intelligent interoperability, security and privacy, human-centric engagement, and sustainable ecosystem design. Each pillar in the ecosystem is interlinked within the AI-driven architecture by ensuring adaptability, seamless integration, and resilience in the healthcare ecosystem. The initial aspect of the framework is related to intelligence through artificial intelligence (AI) in IoMT systems. The focus of this aspect is to enable and provide real-time data exchange, which is not only seamless but also based upon heterogeneous processes. The traditional IoMT systems face significant challenges related to proprietary standards, lack of semantics, consistency, and fragmented data formats. The present framework aims to address such challenges by proposing an artificial intelligence-driven interoperability engine. This aspect is concerned with leveraging natural language processing and machine learning processes to harmonise and standardise healthcare system data. It collaborates with genomic databases, telemedicine platforms, and sensor data, thereby following a unified data fabric approach. This approach enables predictive analytics and holistic patient profiling. Furthermore, the framework also considers dynamic ontology mapping, which ensures semantic interoperability. The role of adaptive communication protocols is duly recognised to ensure real-time responsiveness in healthcare and critical settings, and to optimise data transmission based on edge computing frameworks.

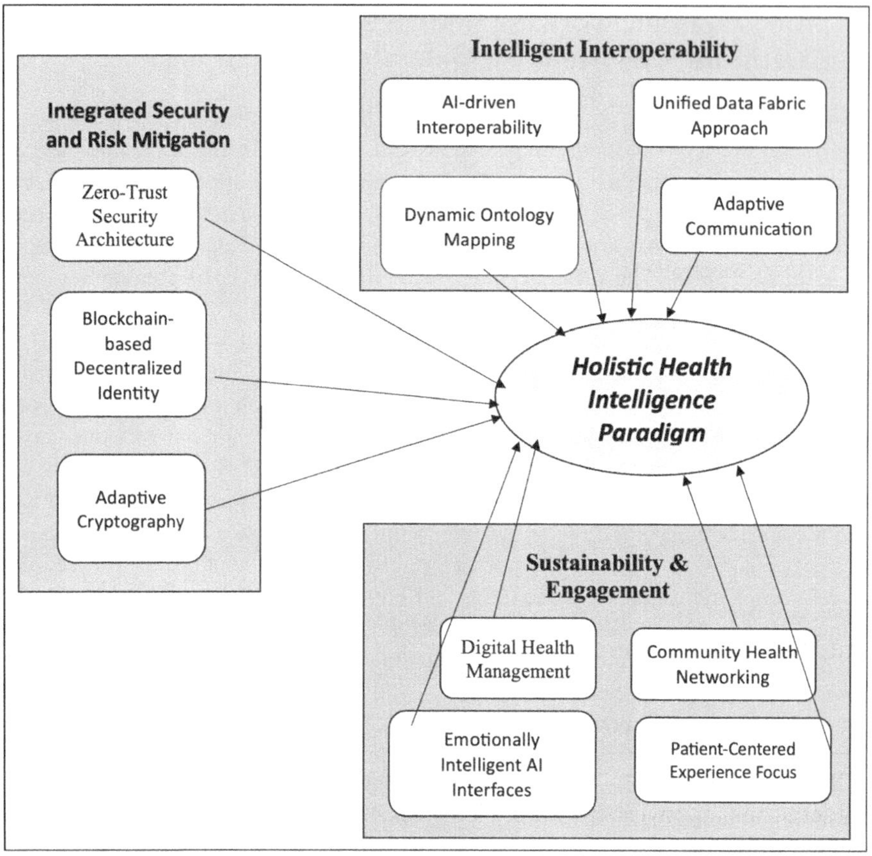

Figure 2: A Framework of Holistic Health Intelligence System in Healthcare 5.0. Source: Authors' creation.

The second aspect of the framework is concerned with ensuring security and privacy with a decentralised governance system. It is pertinent that IoMT healthcare systems shall duly recognise the sensitive nature of health data around increasing cyber threats. It moves beyond static encryption to enable access control mechanisms that are dynamic and based upon utmost risk awareness. The security architecture relies on zero-trust principles and behavioural biometrics for ensuring access to data and systems verification, as well as device integrity. Furthermore, the data access rights are ensured to be controlled by relying on decentralised identification and blockchain systems. Additionally, the emerging vulnerabilities and attack vectors are proposed to be administered using adaptive cryptography. This aspect of the framework aligns with the global protection regulations, thereby ensuring continuous paradigm improvements. The third aspect of the framework is concerned with engaging a human-centric approach through an empathic design. The focus of this approach is different from the one established in the traditional IoMT approach, wherein providing empathetic healthcare experiences to the patients is considered primary in relation to providing only technology-centric solutions. The traditional IoMT healthcare systems mainly relied on providing technological solutions and

Table 2: Main Dimensions of the Holistic Healthcare Intelligence in Healthcare 5.0.

Framework Pillar	Purpose & Focus	Key Challenges Addressed	Innovative Solutions Proposed
Intelligent Interoperability	Enable seamless, real-time data exchange across heterogeneous IoMT systems via AI.	Fragmented data, lack of semantic consistency, proprietary standards.	AI-driven interoperability engine (NLP/ML); dynamic ontology mapping; edge-computing optimized protocols; unified data fabric.
Security & Privacy (Decentralised)	Protect sensitive health data with adaptive, risk-aware, and regulatory-compliant systems.	Static encryption, cyber threats, device vulnerabilities, access control gaps.	Behavioural biometrics; adaptive cryptography; blockchain-based decentralised IDs; GDPR-aligned dynamic controls.
Human-Centric Engagement	Prioritize empathetic, inclusive, and participatory patient & clinician experiences.	Tech-centric designs leading to low adoption, poor usability, health literacy gaps.	Emotionally intelligent AI interfaces (stress detection); participatory co-design; adaptive learning for multilingual/cultural contexts.
Sustainable Ecosystem Design	Minimize environmental impact while ensuring long-term viability of IoMT systems.	High energy use, e-waste, lack of lifecycle accountability in healthcare tech.	Circular health model: energy-harvesting IoMT devices; blockchain asset tracking; carbon-neutral infrastructure; AI-driven resource optimization.

Source: Author's Creation

often overlooked patient needs, leading to poor patient satisfaction and lower adoption rates as highlighted in Table 2. The framework differs from the traditional IoMT healthcare approach and aims to design a participatory methodology that duly involves healthcare professionals and patients in the creation of IoMT-based solutions. Such an approach is not only accessible but also aligns with the real-world healthcare requirements.

The AI-driven interfaces shall be designed in such a manner that they can detect stress levels and are emotionally intelligent to provide the required intervention and personalised healthcare support. The adaptive learning modules aim to extend health literacy across diverse populations by considering different language preferences and cultural contexts. The final aspect of the framework is concerned with developing a sustainable ecosystem design through the mode of circular healthcare models. This aspect is integrated with the third aspect and forms its sub-part. The ecosystem design aims at addressing the increasing concerns of environmental impact and resource inefficiencies in the healthcare systems. This aspect aims to combine the sustainable aspects with sophisticated technological innovation. The energy-efficient IoMT devices are incorporated, aiming to focus on low-power and energy-harvesting systems, which minimise wastage. Furthermore, an asset-tracking system based on blockchain devices is incorporated that suitably monitors the entire life cycle of the IoMT processes right from manufacturing to its disposal, thereby ensuring recycling, reuse, and responsible system management. This aspect also takes into consideration been proposed in the current study, but it is emphasised that they must be operated in integration and with adequate governance. The book chapter has identified a significant gap in the current literature related to the absence of a risk-aware, context-specific security system that adapts suitably to the dynamic nature of IoMT-based devices. The literature gap has been met by inculcating the intricacies concerning zero trust principles, biometrics, and adaptive cryptography, which are embedded in the contextual security pillar of the proposed model of the study.

Another significant finding of the study is related to proposing a framework that is based on a human-centric approach in healthcare systems. The previous studies outlined that traditional IoMT-based approaches in the healthcare system prioritised a technology-centric model, which focuses on functionality and efficiency over the user experience, thereby resulting in suboptimal patient engagement and lower adoption rates. The established literature revealed that there is significant potential to improve healthcare outcomes, and its effectiveness is directly influenced by developing the system, which is based upon a human-centric approach and considers the aspects of accessibility, usability, and alignment with the needs of the patients. The current study highlights the significance of participatory design approaches that duly involve healthcare professionals and patients in the implementation of an IoMT-based healthcare system. The design approach proposed in the current study is empathic in nature as it ensures that the AI-driven healthcare interfaces are emotionally intelligent, can detect stress levels, and are responsive to the patient's behaviour. Such an approach shall not only enhance the quality of the healthcare system but also address the disparities in health literacy across the diverse population.

The book chapter also fosters developing a sustainable approach to an IoMT-based healthcare system. It was found that the healthcare systems are increasingly relying on integrated devices and on the digital infrastructure. Such reliance is raising concerns related to the environmental impact of these technologies. The past literature indicated that IoMT-based healthcare systems are contributing to increased energy consumption and resource inefficiencies. In this regard, the proposed framework of the study focuses on developing a circular health model that amalgamates the sustainability aspect into the core design of IoMT healthcare systems. The study focuses on developing energy-efficient devices and blockchain-based asset tracking through the augmentation of carbon-neutral healthcare infrastructure.

The synthesis of these research findings encouraged building a decentralised governance model that balances stakeholder autonomy with policy oversight. The proposed paradigm aims to leverage smart contracts for the regulatory policy and its enforcement. This shall enable real-time compliance monitoring and streamline healthcare operations. It shall empower collaborative decision-making among the interested stakeholders, which includes regulatory authorities, patients, and healthcare providers. An independent ethics committee proposed in the IoMT architecture shall ensure that the treatment recommendations and the AI-driven diagnostics have the elements of transparency, unbiasedness, accountability, trust, and fairness. The current book chapter aims to fill the crucial voids in the current literature by proposing a well-inclusive framework for IoMT-based healthcare 5.0 systems. The framework emphasises significant aspects of intelligent interoperability, human-centric design, sustainability, and context-aware security in building the next generation of healthcare systems. A theory-driven and multidisciplinary approach sets the path for future research and innovation in this domain and guides policy development among the regulatory authorities. The findings present a strategic roadmap for the healthcare providers who are aiming to offer secure, personalised, future-ready, and adaptable care in an increasingly AI-driven digital landscape of healthcare processes.

6. Implications of the Study

The outcomes of the current book chapter carry noteworthy practical and policy-level implications for the augmentation of IoMT-based processes. The healthcare 5.0 system continues to evolve at a global level and is moving ahead with more intelligent, decentralised, and human-centric models. The technologies must be properly integrated with a suitable paradigm that addresses the concerns related to security management, human engagement, interoperability, and sustainability as outlined in Figure 3. The following are the implications of the study, which set a critical area of impact for conducting future research in this domain.

6.1 Harmonising Interoperability Standards for IoMT System Integration

A major hurdle established in IoMT-based healthcare systems is the lack of semantic and standardised interoperability, which poses significant challenges in the healthcare processes. It is therefore fostered that an AI-driven interoperability framework that utilises natural language processing and machine learning aspects shall be employed for enhancing the cross-platform integration and augmenting its scalability. The establishment of a unified data interface shall interlink all telemedicine platforms, electronic health records, and databases related to genomics. The framework proposed in the current study promotes facilitating real-time clinical decision-making and the conduct of predictive analysis. In this context, the regulatory authorities shall concentrate on the establishment of open standards for enabling seamless healthcare data exchange and maintaining the usability as well as data integrity of IoMT-based healthcare systems.

6.2 Re-envisioning Security and Privacy by Employing Context-Aware Architectures

The two primary critical concerns in IoMT-based healthcare systems are security and privacy of the health information. The present study outlined various limitations of data encryption and access control mechanisms followed in traditional healthcare systems. The proposed security framework in this study has its underpinning in zero trust principles, decentralised identification management, and behavioural biometrics. Such an approach offers an adaptive and risk-aware process for securing health information. These outcomes highlight that the future healthcare security models shall outperform the static protocols and incorporate processes that shall detect threats on a real-time basis and enable behavioural analysis. In this regard, it is suggested that the policymakers and tech developers shall collaborate to administer superior governance frameworks that operate for data auditability, sovereignty, and compliance with global data protection standards.

6.3 Developing a Humanitarian Design as a Facilitator for Patient Involvement

A significant outcome of the present research work is the suggestion of a need to shift from a technology-centric model to human-centric health models. The upsurge in IoMT-based healthcare devices is limited in their approach to achieve widespread adoption due to the issues in usability, lack of empathy, and improper consideration of the patient's needs. The proposed framework of the study is designed on a human-centric approach that focuses on a participatory intention and considers emotionally intelligent AI processes for improving the user experience in healthcare systems. Such a development implies that the policy makers shall consider improving the health literacy, cultural sensitivity, and emotional engagement for improving the trust and long-term adoption of IoMT-based devices. Additionally, the healthcare providers shall work on developing adequate solutions for ensuring that sophisticated technology meets the expectations of the patients.

6.4 Ensuring a Responsible IoMT System Development

The aspect of sustainability in a digital healthcare system is often ignored, and it is highlighted as a matter of huge concern in the established literature concerning contemporary healthcare systems. The current study recognises the urgent need to embed the sustainability concern into the life-cycle design of healthcare IoMT processes. The proposed paradigm of the healthcare model enables adopting energy-efficient devices, carbon-neutral health infrastructure, and blockchain-based asset tracking. Such insights underscore that the future of the healthcare system shall align with the world's sustainability objectives, aiming to reduce digital waste and enable optimisation of energy consumption. In this regard, it is suggested that healthcare professionals should amalgamate environmental impact assessment into their digital procurement strategies to ensure that technological healthcare progress does not degrade ecological structures.

6.5 Advancing Ethical AI Oversight Mechanisms

Integrating AI and decentralised technological systems needs a major overhaul in the way it is governed and regulated in healthcare 5.0 systems. The current study outlines that a decentralised governance model shall be adopted wherein smart contracts are utilised to ensure that decision-making processes are based upon transparency, accountability, and ethics. Such insights shall be applied by the healthcare regulators to ensure that AI-driven diagnosis and treatment processes are unbiased, fair, and explainable. They shall foster the development of an independent ethics committee in a healthcare system to ensure that the AI processes are properly managed, protect human rights, and ensure public trust. The government organisations shall frame

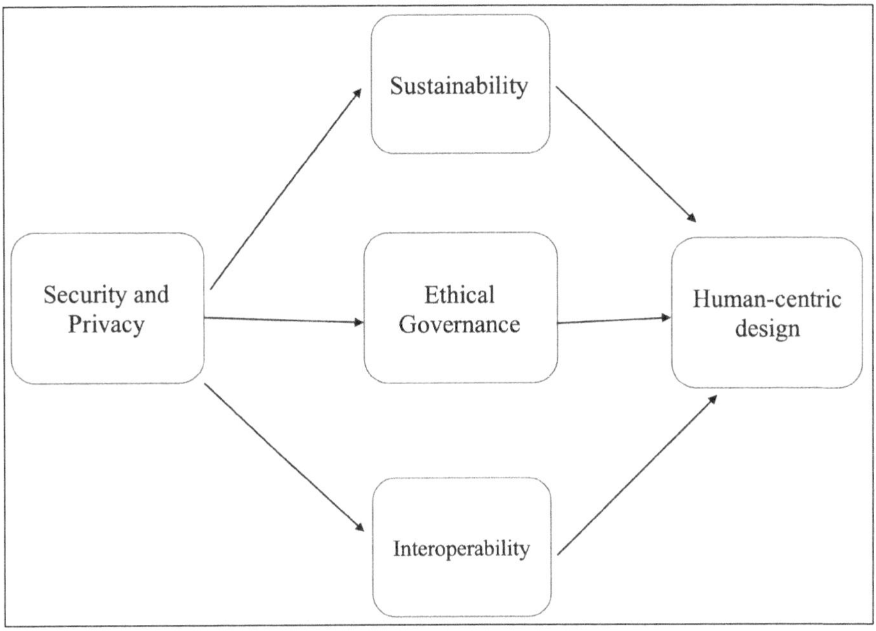

Figure 3: Implications of IoMT-based integrated healthcare 5.0 system. Source: Author's creation.

flexible healthcare regulations to encourage data safety and the upkeep of patient rights.

It is concluded that the above implications comprehensively deal with augmenting technological innovation and influencing healthcare practices in relation to IoMT-based healthcare systems. The study addresses critical gaps concerning interoperability, security management, human engagement, sustainability, and governance. It sets the roadmap for the development of the next-generation healthcare system and provides insights to the policymakers for developing inclusive, intelligent, and secure healthcare technologies.

7. Conclusion and Scope for Future Research Work

The integration of IoMT into the healthcare 5.0 system has presented a transformative leap for developing intelligent, sustainable, and patient-centred healthcare procedures. The present research work outlined a well-inclusive paradigm that extends beyond technological aspects to embrace intelligent interoperability, a well-informed security system, a human-centric model, and an aspect of environmental sustainability. It synthesises prevailing theoretical foundations and builds a holistic framework that addresses the contemporary challenges in the healthcare 5.0 ecosystem. It redefines the role of the Internet of Medical Things in developing a resilient, ethical, and future-ready healthcare. The contribution of the present research work lies in fostering an adaptive and decentralised governance architecture that enables quick responsiveness to evolving clinical, operational, and environmental concerns. Such a framework aligns with the real-world healthcare challenges concerning cybersecurity

Natarajan, R., Lokesh, G. H., Flammini, F., Premkumar, A., Venkatesan, V. K., and Gupta, S. K. (2023). A Novel Framework on Security and Energy Enhancement Based on Internet of Medical Things for Healthcare 5.0. *Infrastructures*, 8(2): 22. https://doi.org/10.3390/infrastructures8020022

Ognjanović, I., Zoulias, E., and Mantas, J. (2024). Progress Achieved, Landmarks, and Future Concerns in Biomedical and Health Informatics. *Healthcare*, 12(20): 2041. https://doi.org/10.3390/healthcare12202041

Prabha, C., Garg, G., and Ahuja, B. (2024). IoT in healthcare: Applications, challenges and future concerns. *2024 OPJU International Technology Conference (OTCON) on Smart Computing for Innovation and Advancement in Industry 4.0*, 1–6. https://doi.org/10.1109/otcon60325.2024.10687721

Pradhan, B., Bhattacharyya, S., and Pal, K. (2021). IoT-Based Applications in Healthcare Devices. *Journal of Healthcare Engineering*, 2021, 1–18. https://doi.org/10.1155/2021/6632599

Rahman, A., Kundu, D., Debnath, T., Rahman, M., Das, U. K., Miah, A. S. M. et al. (2025). From AI to the Era of Explainable AI in Healthcare 5.0: Current State and Future Outlook. *Expert Systems*, 42(6). https://doi.org/10.1111/exsy.70060

Rashid, S., and Nemati, A. (2024). Human-centered IoT-based health monitoring in the Healthcare 5.0 era: Literature descriptive analysis and future research guidelines. *Discover Internet of Things*, 4(1). https://doi.org/10.1007/s43926-024-00082-5

Riaz, M. Z., Behera, B. K., Mumtaz, S., Al-Kuwari, S., and Farouk, A. (2025). Quantum machine learning for energy-efficient 5G-enabled IoMT healthcare systems: Enhancing data security and processing. *IEEE Internet of Things Journal*, 1–1. https://doi.org/10.1109/jiot.2025.3586090

Sadeghi, M., and Mahmoudi, A. (2024). Synergy between blockchain technology and internet of medical things in healthcare: A way to sustainable society. *Information Sciences*, 660: 120049. https://doi.org/10.1016/j.ins.2023.120049

Samadhiya, A., Kumar, A., Arturo Garza-Reyes, J., Luthra, S., and Del Olmo García, F. (2024). Unlock the potential: Unveiling the untapped possibilities of blockchain technology in revolutionizing Internet of medical things-based environments through systematic review and future research propositions. *Information Sciences*, 661: 120140. https://doi.org/10.1016/j.ins.2024.120140

Sharma, B., Kaushal, D., Sharma, M., Joshi, S., Gopal, S., and Gupta, P. (2023). Integration of AI, digital twin and Internet of Medical Things (IoMT) For healthcare 5.0: A bibliometric analysis. *2023 International Conference on Advances in Computation, Communication and Information Technology (ICAICCIT)*, 1313–1318. https://doi.org/10.1109/icaiccit60255.2023.10466141

Shukla, S., Thakur, S., Hussain, S., and Breslin, J. G. (2021). Identification and authentication in healthcare Internet-of-Things Using integrated fog computing based blockchain model. *Internet of Things*, 15: 100422. https://doi.org/10.1016/j.iot.2021.100422

Singh, M. (2025). Enhancing Site Reliability Engineering Through AIOps: A Framework for Next-Generation IT Operations. *Asian Journal of Research in Computer Science*, 18(4): 272–284. https://doi.org/10.9734/ajrcos/2025/v18i4619

Ullah, S., Niu, B., and Meo, M. S. (2025). Advancing sustainability in China by uncovering the role of eco-innovation, sustainable urbanization and financial development: Evidence from novel wavelet approaches. *International Journal of Sustainable Development & World Ecology*, 32(1): 110–125. https://doi.org/10.1080/13504509.2024.2412153

Wazid, M., Das, A. K., Rodrigues, J. J. P. C., Shetty, S., and Park, Y. (2019). IoMT malware detection approaches: analysis and research challenges. *IEEE Access*, 7: 182459–182476. https://doi.org/10.1109/access.2019.2960412

CHAPTER 2

Challenges of Cybersecurity in Intelligent Health Systems

Saurabh Singhal,[1] *Ajeet Kumar Sharma*[2,*] and *Vivek Kumar Singh*[3]

1. Introduction

New technologies are emerging in the health sector, which is moving this field forward. Various technologies like AI, IoT, ML, and edge computing are being used in digital healthcare, which is enhancing patient outcome and experience.

By using these technologies, the medical field has been enhanced, but the risk of cybersecurity has also increased. Digital health records contain confidential data of patients, causing a risk of their confidential information being disclosed (Mimmo 2025). Confidentiality, integrity, and availability of health-related records are compromised due to cyberattacks. Healthcare organizations utilize electronic health records, telemedicine, and remote patient monitoring, which increases the risk of security issues. Figure 1 shows the overview of intelligent health system.

Healthcare information is very sensitive; any risk of cybersecurity incidents can directly put human lives at risk. Ransomware attacks can encrypt the data and make it inaccessible to stakeholders. Various popular attacks, such as WannaCry, highlighted the need for strong cybersecurity measures in healthcare. Healthcare systems are governed by a certain set of rules, ethics, and laws such as HIPAA and GDPR to handle, store, and share patient information. Various security-related challenges should be addressed, but handling these is not an easy task; it requires a multilayered approach. There is a need for an environment that ensures the security of devices.

[1] Department of CSE, Greater Noida Institute of Technology (Engineering Institute), Greater Noida, Uttar Pradesh, India.
Email: me.ssaurabh@gmail.com
[2] School of Computer Science & Engineering, IILM University, Greater Noida, Uttar Pradesh, India.
[3] Department of Computer Science & Engineering, Sharda University, Greater Noida, Uttar Pradesh, India.
Email: singh.vivekrajput@gmail.com
* Corresponding author: ajeetsharma1989@gmail.com

DOI: 10.1201/9781003783855-3

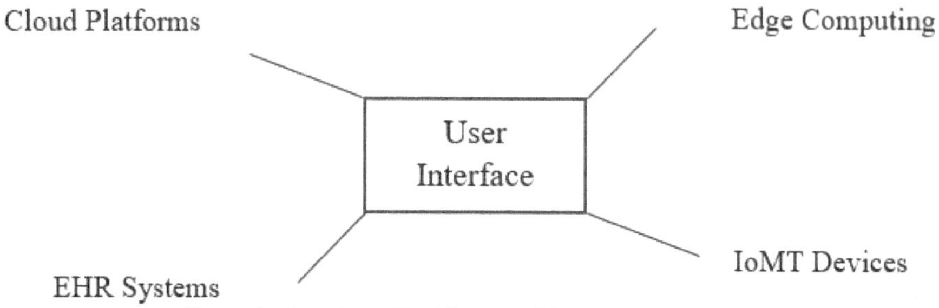

Figure 1: Overview of Intelligent Health System Architecture

The major concern exists regarding the extent to which AI-enabled systems should be applied to make decisions and whether humans should remain in the loop. The moral and ethical concerns also involve the impact of this on employment and cultural values (Mimmo 2025).

1.1 Motivation

Every manufacturer uses its own configuration of devices, data pattern and protocols, definitions of devices, data format, and communication protocols. Various communication protocols have been utilized in IoMT, whose performance depends on the functionalities they generally offer. An ideal IoMT platform must be developed in a way, where it can use existing protocols and also adapt to new ones. Electronic health records (EHR) are being used to store patient data in hospital information systems. The complexity and large amount of data are creating several challenges to deal with. The diversity in health information systems, such as third-party applications, medical equipment, and interoperability with less disruption, also increases the possibilities of cyberattacks (Rubí et al. 2019).

Attackers take advantage of weaknesses, as most organizations resist or postpone patching to ensure availability. Low cybersecurity awareness on the part of healthcare personnel, sometimes called the “weakest link,” helps make social engineering attacks successful. Further, most hospitals suffer from a lack of cybersecurity experts, which results in overburdened IT security teams. Growing usage of Bring Your Own Device (BYOD) policies and remote work arrangements complicates endpoint security and device management further. There can be various entry points for cyberattacks in a typical healthcare environment, as depicted in Figure 2.

Smart health systems must balance privacy and utility of data to ensure ethical practices. AI models can help, but their wholesale collection raises concerns about informed consent, data ownership, and secondary use (Reza et al. 2020). Certain groups, like children, the elderly, and those with disabilities, may be more susceptible to data misuse. Equitable cybersecurity models must ensure patient autonomy, equity, and accountability. Blockchain technology can enhance transparency and trust in sharing medical data by establishing immutable audit trails and cryptographically safe transactions. Federated learning is being explored for AI models on distributed datasets without patient raw data sharing, but technical, legal, and infrastructural

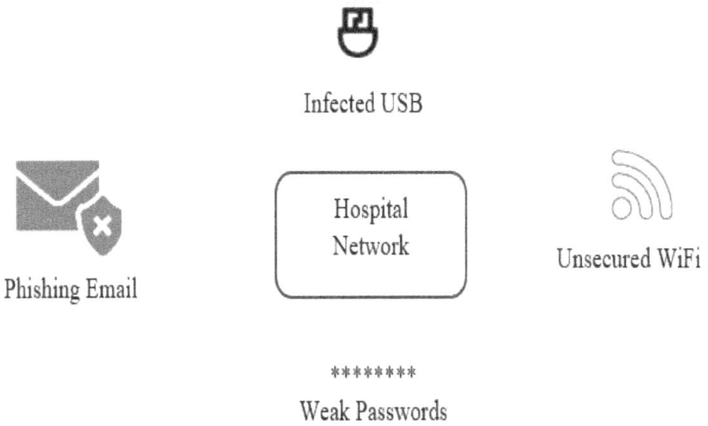

Figure 2: Cyberattack Entry Points in a Typical Healthcare Environment.

challenges hinder these methods. A "Security-by-Design" approach is gaining traction in healthcare IT development, emphasizing secure coding, *in situ* threat modeling, and penetration testing. Zero Trust architectures can mitigate lateral movement during attacks, and Intrusion Detection and Prevention Systems (IDPS) can track anomalous activity and react to incidents in real time. Cybersecurity in smart health systems is an ongoing process with dynamic threats, and healthcare institutions must maintain a dynamic cybersecurity posture that includes threat intelligence, behavior analytics, and frequent audits (Thangalakshmi et al. 2025).

1.2 Contribution

This chapter makes the following contributions on cybersecurity in intelligent health systems. A structured overview of intelligent health systems, detailing the architecture and layered design to understand the technologies in healthcare. It also highlights the role of the Internet of Medical Things (IoMT), its applications, and security vulnerabilities in the environments. The significant contributions of cloud and edge computing are also discussed. The chapter focuses on how artificial intelligence and machine learning contribute to automating security monitoring. The systematic study of cybersecurity, the threat landscape, and common attack vectors is also included.

2. Intelligent Health Systems: Architecture and Key Components

The integration of digital technologies in healthcare has created an intelligent health system, which is based on interconnected devices, decision-making, personalized medicine, and automation. Deep learning can be utilized to collect detailed information of patients suffering from serious diseases and can also be used to analyze the medical data obtained from various smart devices. The real-time monitoring of health data prevents various chronic diseases and helps maintain a healthy lifestyle (Sujith et al. 2022). The design of the system should provide scalability, strength, and security to the digital health information system.

Table 1: Core Components of Intelligent Health Systems and Their Functions.

Component	Function in IHS	Challenges
IoMT Devices	Real-time health data acquisition	Device security, interoperability, battery life
EHR Systems	Centralized patient data management	Privacy, legacy system integration
Cloud Computing	Scalable storage and remote processing	Latency, data jurisdiction, cost
Edge Computing	Low-latency local computation	Resource constraints, device heterogeneity
Artificial Intelligence	Predictive modeling and decision support	Bias, explainability, data dependency
Big Data Analytics	Insights from heterogeneous data	Data quality, infrastructure requirements
Communication Layer	Secure data transmission	Bandwidth, encryption, protocol fragmentation
Human-Computer Interfaces	Interaction with clinicians and patients	Usability, alert fatigue, accessibility

3.1 Dynamic Threat Surface in Smart Healthcare

a. Internet of Medical Things (IoMT) aims to improve the patient's quality of life by personalized health services without the limitations on time and location (Papaioannou et al. 2022). The key reasons to explore the threat surface are outlined below:
b. Growing IoMT Devices: Each connected device is at risk of threats or unauthorized access.
c. Data Interoperability & Mobility: Interoperability of patient data between multiple platforms and institutions creates risks of tampering.
d. BYOD Policies: Use of personal devices by employed medical personnel allows uncontrolled entries.
e. Cloud and Edge Integration: Although enhancing latency and scalability, they also create new vectors of attack.
f. Legacy Systems: Outdated systems without current security patches remain vulnerable.

3.2 Categorization of Cyber Threats in Intelligent Health Systems

Variants of malware, such as ransomware, spyware, and Trojans, are frequently utilized to interfere with or hijack health systems. In ransomware attacks, high-priority patient information is encrypted and taken hostage, with the common outcome being suspended services and financial losses (Kulkov et al. 2023).

a. Phishing: Phishing emails and SMS target human error to obtain access credentials or deliver malicious payloads. A spear-phishing email that used a hospital administrator persona deceived staff into providing login credentials, allowing attackers remote access to healthcare records (Ani et al. 2019).
b. Denial of Service (DoS) and Distributed DoS (DDoS): These assaults swamp systems with traffic, making medical services such as teleconsultation or emergency alert systems unavailable when they are most needed.
c. Insider Threats: Staff or contractors with approved access can deliberately or unwittingly leak or manipulate patient data.
d. Data Breaches and Identity Theft: Unauthorized access to medical databases can result in mass-scale theft of sensitive health information, which is sold on the dark web or used for identity fraud.
e. Supply Chain Attacks: Attackers target third-party vendors (i.e., software firms, device makers), introducing vulnerabilities into otherwise secure settings.
f. AI-Specific Attacks: As AI becomes integral in diagnostics and decision support, machine learning models may be targeted by adversarial attacks, leading to misdiagnosis or biased outcomes.

3.3 Common Attack Vectors and Entry Points

Cybercriminals use a variety of entry points to attack healthcare networks. Table 2 presents the most common attack vectors and their impact. Table 2 shows the common cyberattack vectors in intelligent health systems, while Table 3 summarizes the cybersecurity risks of the key technologies in HIS.

Table 2: Common Cyberattack Vectors in Intelligent Health Systems.

Attack Vector	Description	Example Impact
Phishing	Email/SMS tricking users into revealing credentials or clicking malicious links	Compromise of EHR access credentials
Infected USB Devices	Malware spread through physical device insertion	Introduction of ransomware in hospital PCs
Weak Passwords	Easy-to-guess or reused credentials enable brute force attacks	Unauthorized access to patient monitoring systems
Unpatched Software	Exploits targeting known vulnerabilities in outdated applications	Full system compromise
Third-Party Applications	Vulnerable APIs or libraries in health apps	Data exfiltration through mobile platforms
Public Wi-Fi & BYOD	Insecure devices or networks bypass organizational security	Man-in-the-middle attacks

Table 3: Key Technologies and Their Associated Cybersecurity Risks in IHS.

Technology	Role in HIS	Primary Cybersecurity Risks
IoMT Devices	Real-time patient monitoring	Device hijacking, data spoofing, unauthorized access
Cloud Computing	Data storage and analytics	Data breaches, insecure APIs, lack of data control
Edge Computing	Low-latency processing near patient	Physical tampering, firmware attacks
Artificial Intelligence	Diagnostic and predictive modeling	Model poisoning, adversarial inputs, decision integrity
EHR Systems	Centralized patient data	Data leakage, improper access, ransomware
Telemedicine Platforms	Remote consultations	Video interception, phishing, authentication flaws

4. Data Privacy and Protection Challenges in Intelligent Health Systems

The incorporation of cutting-edge digital technologies in healthcare has brought unprecedented access to patient information and real-time medical care. From cloud-based electronic health records (EHRs) and AI-driven diagnostics to wearable IoMT devices, smart health systems have extensive dependence on the constant stream of sensitive data (Freedman et al. 2013). Although this change enhances care provision, it also creates fundamental data privacy and protection issues. Guaranteeing the confidentiality, integrity, and availability of healthcare data in such highly interconnected settings is a key concern for regulators, providers, and patients alike. This section discusses the special data protection challenges in smart healthcare, highlights main vulnerabilities, presents relevant regulations, and offers current best practices and limitations.

4.1 Principal Privacy and Protection Challenges

Smart healthcare systems face various challenges in maintaining privacy and security. Traditional techniques are not sufficient in terms of transparency and accountability. These challenges arise because patients are often unaware of their data's location, access, and usage. AI models trained on health data often lack explainability. Third-party and supply chain risks arise from routine data transfers, increasing the need for end-to-end security and compliance with regulations. Cross-border data transfers, particularly in cloud computing platforms, increase jurisdictional and regulatory challenges, necessitating compliance with local laws like GDPR or HIPAA. Additionally, anonymized data can be reverse-engineered by AI, potentially re-identifying individuals. Table 4 shows major data privacy challenges being faced by smart healthcare.

Table 4: Major Data Privacy Challenges in Smart Healthcare.

Challenge	Description	Implications
Data Proliferation	Data generated/stored in diverse systems	Difficult to enforce universal protection
Weak Consent Mechanisms	Ambiguous or bundled consent for health apps	Loss of control over data
Re-identification Risk	AI can match anonymized data with external sources	Violation of patient anonymity
Shadow Health Records	Health data gathered outside of traditional healthcare (e.g., fitness apps)	No regulatory oversight
Third-party Access	Vendors may mishandle or misuse patient data	Breaches and data leaks

4.2 Technical Data Protection Challenges

AI-driven healthcare presents several privacy and security challenges. Encryption overhead, key management, secure interoperability, data provenance and integrity, and privacy issues are all significant challenges. Encryption ensures confidentiality but adds latency and computational overhead, making it undesirable in real-time applications like ICU monitors. Key storage and exchange are challenging, especially in distributed and mobile applications. Data integrity is critical for forensic and compliance reasons. AI may also have some privacy issues, such as bias, disclosure of training data, and model inversion risks. Table 5 summarizes the different techniques used for data protection along with their challenges.

Table 5: Techniques for technical data protection with challenges.

Technique	Description	Challenges
Data Anonymization	Removing identifiers like names or IDs	Risk of re-identification
Differential Privacy	Adds noise to data to prevent individual identification	Accuracy vs privacy trade-off
Homomorphic Encryption	Enables computation on encrypted data	Computationally expensive
Federated Learning	AI training without centralizing data	Communication overhead; model poisoning risk
Blockchain for Audit Trails	Immutable logs of data access	Scalability and energy concerns

on digital infrastructure, guaranteeing that technology is morally aligned and legally compliant is no longer a choice—it is a necessity.

This section discusses the intersection of law and ethics in the field of cybersecurity for intelligent health systems, with emphasis on patient rights, informed consent, data governance, algorithmic accountability, cross-border regulation, and compliance standards.

a. One of the cornerstone moral principles in health care is respect for patient autonomy, meaning patients should maintain complete authority over their own personal health data. With intelligent health systems, this principle is problematic. Contemporary systems gather information continuously—often incidentally—via wearable sensors, mobile health applications, and remote monitors. Patients are not always aware of what information is gathered, how it is preserved, and to whom it is transmitted. Patients should be informed about the extent of data collection, processing purposes, third-party sharing risks, and rights of access, rectification, or erasure, extending beyond paper signatures. Presenting this information in a clear and accessible way to elderly or less digitally savvy populations is still a major challenge. Additionally, old-fashioned ethical consent fails where AI systems operate through constant learning and bundling of data from a variety of sources.
b. Ethically, privacy of the patient is central to the preservation of trust in healthcare professionals. Violations of confidentiality not only contravene ethical standards but also may cause tangible harm like identity theft, social stigma, or psychological trauma. Patient information stored in cloud servers across borders, transmitted to insurance providers or third-party analytics companies, and used for AI model training without anonymization increases risk, as trust can be lost when patients perceive monitoring or feel that information is being sold without their clear consent. This might deter patients from providing vital health data or engaging actively with digital care systems, ultimately compromising care quality (Putra et al. 2024).

8. Strategies for the Strengthening of Cybersecurity in Smart Healthcare

Smart healthcare infrastructures, defined by connected devices, cloud-based systems, and artificial intelligence-enabled analytics, have become more susceptible to cyberattacks. These attacks not only threaten patient safety and privacy but also interfere with clinical workflows, lower confidence in digital innovation, and burden healthcare institutions with financial expenditures. Hence, the development of a strong cybersecurity infrastructure is essential to supporting the integrity, availability, and confidentiality of healthcare services. This part describes a multilayered approach for smart healthcare cybersecurity improvement, including technical tools, policy initiatives, organizational routines, and human-centered activities.

8.1 Conventional Perimeter-Based Security

The conventional perimeter-based security paradigm—with systems trusting internal devices and users by default—is no longer viable in a health environment characterized by mobility, remote access, and device diversity. Zero Trust Architecture (ZTA) takes a stance in presuming no user or device should be trusted by default, either inside or outside the network. ZTA's key principles include least-privilege access, continuous authentication through Multi-Factor Authentication (MFA), micro-segmentation to restrict threats' lateral movement, and device and user verification using endpoint detection tools and behavioral analytics. By implementing ZTA, hospitals and clinics can drastically reduce the blast radius of attacks such as ransomware or insider threats.

8.2 Identity Management

Identity and Access Management (IAM) solutions provide security so that only authorized staff members have access to sensitive data and systems. With the variety of functions in healthcare—physicians and nurses, third-party technicians—there is a need for strong IAM policies. Implementing effective Information Access Management (IAM) practices such as Role-Based Access Control (RBAC), Multi-Factor Authentication (MFA), Single Sign-On (SSO), and Access Auditing can help prevent unauthorized access to data, insider attacks, and compromised credentials, which are major causes of healthcare data breaches.

Intelligent healthcare systems communicate data between IoT devices, mobile applications, EHR systems, and cloud storage, necessitating end-to-end encryption. Methods include at-rest encryption using AES-256, in-transit encryption using TLS and HTTPS, and blockchain-based integrity checks to maintain immutability of patient records. These practices enhance security and comply with HIPAA and GDPR requirements.

Cyberattacks often exploit outdated software or unprotected vulnerabilities, particularly in hospitals where legacy medical devices often lack routine updates. Mitigation strategies include automated patch management, vulnerability scanning, and system hardening. These measures minimize attack surface and remove low-hanging fruit for cybercriminals, ensuring the safety of all devices (Raj et al. 2024).

9. Future Directions and Research Challenges in Cybersecurity for Smart Healthcare

Healthcare systems continue their digital transformation journey by integrating Internet of Medical Things (IoMT), electronic health records (EHR), cloud computing, artificial intelligence (AI), and 5G. The complexity and cyber threat landscape have expanded rapidly. The increasing data volumes, decentralized delivery, and AI-enabled decision-making may create new scopes of vulnerability if not handled carefully. Although current strategies provide a solid foundation, they are still not sufficient for future healthcare environments of complicated networks. This section explores the future directions and research challenges that must be addressed to secure intelligent health systems over the next decade. Federated learning (FL) has emerged as a promising paradigm for training AI models collaboratively across decentralized data sources (e.g., multiple hospitals or clinics) without sharing raw data. This is crucial for privacy preservation, regulatory compliance (e.g., HIPAA, GDPR), and protecting patient confidentiality.

9.1 Real-Time Threat Detection and Blockchain

The proliferation of IoMT devices demands real-time threat detection close to the data source. Traditional centralized systems introduce latency and potential single points of failure. Edge AI, which pushes analytics and anomaly detection to local nodes (e.g., gateways, edge servers), can address this need. Edge devices face challenges in complex models, heterogeneous firmware and operating systems, physical and firmware vulnerabilities, and difficulty in model update management and patching. Future opportunities include lightweight deep learning models for edge processors, federated edge AI for decentralized intelligence and data protection, and adaptive security orchestration frameworks with edge–cloud coordination.

Blockchain technology holds promise for tamper-proof, auditable, and decentralized data exchange across healthcare institutions. Smart contracts can enforce access policies, while immutable ledgers enhance data integrity and provenance. Public blockchains face limitations in scalability due to transaction delays, storage burden, legal acceptance, and interoperability with existing systems. Research focuses on lightweight healthcare blockchain protocols, hybrid architectures combining off-chain storage and on-chain verification, and GDPR-compliant blockchain models addressing the "right to be forgotten."

9.2 Post-Quantum Cryptography for Healthcare Security

Quantum computing threatens the foundations of modern cryptography. Algorithms like RSA and ECC may become vulnerable to quantum attacks, making post-quantum cryptography (PQC) an urgent area of research in securing healthcare communications and identity systems. Challenges in healthcare include transitioning legacy systems to quantum-resistant algorithms, balancing computational overhead with device usability, and ensuring regulatory recognition and standardization. Future scope includes integrating lattice-based and hash-based encryption in medical IoT protocols, developing hybrid encryption models during transition phases, and creating quantum-safe identity management and digital signature schemes (Saberi et al. 2024).

10. Conclusion

The emergence of smart healthcare systems has brought efficiency and patient care benefits but also makes healthcare vulnerable to advanced cybersecurity threats such as ransomware and adversarial attacks on AI-driven diagnostic systems. In adopting technologies such as IoMT, cloud computing, AI, and edge computing, healthcare organizations need to counter new attack surfaces and security risks.

Major cybersecurity issues comprise data breaches, unauthorized access, vulnerability of medical devices, and exploitation of AI-based healthcare platforms. Strong authentication, encryption, secure communication protocols, and real-time threat detection are essential for ensuring the confidentiality, integrity, and availability of sensitive medical information. Smart health ecosystems of the future need adaptive, context-sensitive, and proactive cybersecurity measures such as privacy-protecting AI models, secure data-sharing technologies like blockchain, post-quantum cryptographic protocols, and robust architectures for autonomous threat detection and response. Future security strategies need to take into account the ethical and regulatory aspects of healthcare, merging patient consent, transparency, and data governance with security approaches. Cybersecurity integrated with healthcare is essential for a safe, intelligent healthcare environment that calls for cross-professional training, ongoing education, and the development of cybersecurity-conscious clinicians and healthcare IT professionals.

References

Ani, U. D., He, H., and Tiwari, A. (2019). Human factor security: evaluating the cybersecurity capacity of the industrial workforce. *Journal of Systems and Information Technology*, 21(1): 2–35.

Ajeigbe, K., and Wallace, F. (2024). Interoperability and Cybersecurity: Ethical Responsibilities of Healthcare Providers.

Ayala, L. (2016). How hackers gain access to a healthcare facility or hospital network. In *Cybersecurity for Hospitals and Healthcare Facilities: A Guide to Detection and Prevention* (pp. 9–18). Berkeley, CA: Apress.

Fernandez, E. B., and Brazhuk, A. (2024). A critical analysis of Zero Trust Architecture (ZTA). *Computer Standards & Interfaces*, 89: 103832.

Freedman, A. M., Mindlin, M., Morley, C., Griffin, M., Wooten, W., and Miner, K. (2013). Addressing the gap between public health emergency planning and incident response: Lessons learned from the 2009 H1N1 outbreak in San Diego County. *Disaster Health*, 1(1): 13–20.

Gokilakrishnan, G., Vigneshkumar, M., Dhanamurugan, A., Bhasha, A., Subbiah, R., and Anandakumar, H. (2023, March). A review of applications, enabling technologies, growth challenges and solutions for IoT/IIoT. In *2023 9th International Conference on Advanced Computing and Communication Systems (ICACCS)* (Vol. 1, pp. 2241–2250). IEEE.

Hanna, M. G., Pantanowitz, L., Jackson, B., Palmer, O., Visweswaran, S., Pantanowitz, J. et al. (2025). Ethical and bias considerations in artificial intelligence/machine learning. *Modern Pathology*, 38(3): 100686.

Indu, I., Anand, P. R., and Bhaskar, V. (2018). Identity and access management in cloud environment: Mechanisms and challenges. *Engineering Science and Technology, an International Journal*, 21(4): 574–588.

Kulkov, I., Tsvetkova, A., and Ivanova-Gongne, M. (2023). Identifying institutional barriers when implementing new technologies in the healthcare industry. *European Journal of Innovation Management*, 26(4): 909–932.

Lehto, M., Neittaanmäki, P., Pöyhönen, J., and Hummelholm, A. (2022). Cyber security in healthcare systems. In Cyber Security: Critical Infrastructure Protection (pp. 183–215). Cham: Springer International Publishing.

Mimmo, S. S. (2025). Engineering intelligent health systems: ai-powered business analytics for informed clinical decision-making. *Pacific Journal of Advanced Engineering Innovations*, 2(1): 4–11.

Naithani, K., Raiwani, Y. P., Tiwari, S., and Chauhan, A. S. (2024). Artificial intelligence techniques based on federated learning in smart healthcare. In *Federated Learning for Smart Communication Using IoT Application* (pp. 81–108). Chapman and Hall/CRC.

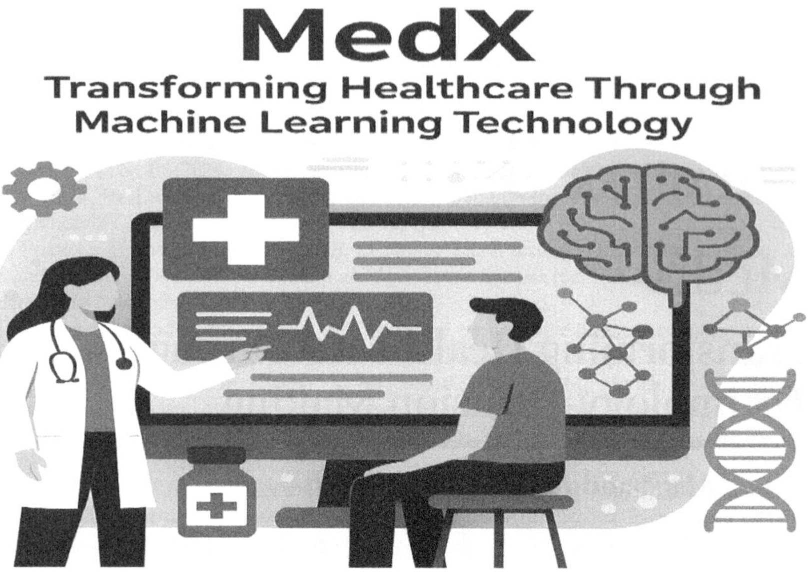

Figure 1: MedX: Transforming Healthcare Through Machine Learning and GAN Technology.

of patients have experienced delays in care due to such problems (WHO 2022). To overcome these challenges, innovative solutions are needed to integrate patient information, improve data flow, and enable proactive healthcare (El Sherif et al. 2022). Leveraging AI alongside cloud-based systems and mobile health platforms can improve patient engagement, facilitate early disease detection, and support a more connected healthcare network. The paper by Perez et al. (2025) introduces MedX, an AI-driven healthcare application that unifies medical data, predicts diseases, provides real-time environmental inMedX uses three key AI algorithms Random Forest for accurate disease prediction, K-Nearest Neighbors (KNN) for BMI-based diet recommendations, and Conditional Generative Adversarial Networks (CGAN) for forecasting environmental health RISK SIGHTS, supports BMI-based health planning, and offers teleconsultations (Bertl et al. 2024). MedX aims to deliver a patient-focused, data-driven system that enhances care quality, accessibility, and efficiency across healthcare environments (Neurocomputing 2025).

2. Motivation

MedX drives healthcare transformation by merging AI analytics with intuitive design, improving access, engagement, and proactive care while addressing privacy, scalability, and provider adoption challenges, building trust for a smarter, resilient healthcare future (Kumar and Singh 2023).

3. Literature Survey: Challenges in the Healthcare Sector

Even with the rapid growth of technology, healthcare still faces many challenges when using AI-based systems (Gupta and Lee 2023). One major issue is the lack of connected and complete patient records. Medical data is often stored in different formats and systems, making it hard for doctors to share and use the information effectively (Chen et al. 2023b). Poor-quality data, missing details, and bias in the datasets can also reduce the accuracy of AI models, especially when the data comes from a small or limited group of patients (Sharma and Patel 2023). Privacy and security are another big concern because health information is highly sensitive and must be protected while still being useful for medical care and research (Chen and Wang 2023a). In many cases, AI systems are not fully tested in different hospitals or regions, and they are not always easy to fit into the daily work of healthcare staff. Survey of Explainable AI (2024) Doctors also need clear explanations of how AI systems make decisions so they can trust and act on the results. In addition, people living in rural or low-income areas may not have good internet access, digital skills, or devices, which limits their ability to benefit from AI tools (Miller et al. 2022). Running these systems can also be expensive, especially for advanced imaging and large AI models that need powerful computers (Rahimi et al. 2024, Infermedica 2023). Research from 2023 to 2025 has studied these problems in detail. For example, Perez et al. (2025) reviewed the use of AI and telemedicine in rural healthcare and pointed out the challenges with infrastructure and data security. Bertl et al. (2024) discussed the main issues in bringing AI into healthcare, such as privacy rules and data integration. Other studies in *Neurocomputing* (2025) and *Medical Image Analysis* (2025) explored technical limits in improving and translating medical images (Siira et al. 2025). Chen et al. (2023b) surveyed in the importance of making AI results easy for doctors to understand. The current MedX system also has its own limits. It uses past data from one region, has not yet been tested widely in real clinical settings, and only partly supports data sharing between systems as shown in Table 1. Its explainability tools are still basic, security upgrades are in progress, and there are still challenges in reaching rural patients and reducing system costs.

3.1 Technological Gaps and Global Trends

While Electronic Health Records (EHRs) are widespread globally, their utility is restricted by interoperability issues, especially in low-resource settings (Mohsen et al. 2022). Countries like the US, UK, and Germany report improved health outcomes due to EHR adoption, but challenges in cost-effectiveness and real-time data integration persist (MDPI 2024). Mobile platforms such as India have improved access to healthcare but still lack AI-driven analytics for early detection.

3.2 Role of AI in Healthcare

AI and machine learning have revolutionized healthcare diagnostics and patient monitoring. Multi-modal AI models combining clinical notes, EHR data, and imaging show significant improvements in diagnostic precision (Kerala e-Sanjeevani n.d.). However, concerns around algorithmic bias, data fairness, and clinical explainability remain significant. NLP-based triage and AI chatbots are under development but are rarely deployed at scale due to data privacy, regulatory, and usability constraints (Smith et al. 2022).

Table 1: Challenges in the Healthcare Sector.

Study (Year)	Focus	Key Findings	Relevance to MedX
Perez et al. 2025	Systematic review of AI + telemedicine in rural care	AI improves diagnostic support; telehealth expands access, but barriers include infrastructure, literacy, and privacy needs.	Supports MedX teleconsultation and remote monitoring modules; highlights equity and security guardrails.
Bertl et al. 2024	Challenges for AI in healthcare systems	Summarizes challenges across data, decision support, software engineering, regulation, and privacy-preserving ML.	Frames MedX architecture choices (interoperability, safety cases, privacy).
Deep learning for medical imaging super-resolution 2025	SR in medical imaging	Reviews DL-SR methods, noise/artefacts, evaluation, and deployment hurdles for higher-quality images.	Informs MedX's image enhancement pipeline and QA metrics.
Medical image translation with deep learning 2025	Image-to-image translation	Surveys advances and datasets; stresses validation and domain-shift risks for clinical use.	Guides MedX's imaging modules (e.g., modality harmonization) and external validation plans.
Survey of explainable AI in healthcare 2024	XAI for clinical models	Reviews XAI techniques and open problems; calls for human-centered, task-grounded explanations.	Motivates MedX's clinician-facing explanations, uncertainty display, and audit trails.

3.3 Accessibility and User-Centered Design

A major challenge in deploying digital health solutions is accessibility, as digital tools are often not inclusive of users with limited technical literacy (Frontiers 2025). Studies emphasize the importance of intuitive design for ensuring wide adoption. Usability gaps and poor design increase dropout rates, particularly among older adults or rural populations.

3.4 Gaps in Existing Systems

Despite widespread EHR and telehealth adoption, integration remains a major issue as discussed in Table 2. Most systems function in silos and are limited in scalability, personalization, and predictive insights. MedX seeks to address these gaps by combining AI disease prediction, real-time environmental data, and user-friendly interfaces into one cohesive system.

Table 2: Existing Work in Healthcare Digitalization.

Study/Source	**Approach**	**Limitations**	**Accuracy Metrics**
Miller (2022)	Fragmented EHR analysis	Lack of data interoperability	—
El Sherif et al. (2022)	AI tools in telehealth during COVID-19	Data security, low real-world validation	—
Chen et al. (2023)	Bias in EHR-based AI models	Lack of fairness in outcomes	N/A
Mohsen et al. (2022)	Multimodal AI (EHR + Imaging)	Needs high computing resources	> 90% accuracy
Siira et al. (2025)	NLP in clinical notes	Only 7% clinically deployed	—
Rahimi et al. (2024)	AI for clinical communication	Needs evaluation in real settings	—
Frontiers (2025)	Chatbots for triage	Lacks integration and personalization	—
Brown et al. (2020)	User-centered design in healthcare	Usability issues in rural tools	—
MDPI (2024)	Rural AI/Telehealth review	Poor infrastructure and literacy	—
Smith et al. (2022)	ML for chronic disease management	Dataset limitations	—
Kerala e-Sanjeevani (2003)	Mobile telehealth	Lacks AI or personalization	—
Amjad et al. (2023)	AI in diagnostics	Limited regulatory support	—
Wired (2023)	Clinical validation of health apps	Only 20% meet clinical standards	—
Early Sense (2022)	Mattress sensors for vitals	Cost and implementation challenges	90% + accuracy
Heidi Health (2025)	AI scribes	Integration issues	—

4. Research Gap

Despite advancements in AI, EHRs, and telemedicine, healthcare systems still lack:

- Seamless integration of diverse data types (clinical notes, imaging, environmental data)
- User-friendly platforms designed for both clinicians and patients
- Scalable solutions for rural or under-resourced regions
- Real-time predictive analytics
- Interoperable frameworks for cross-platform data sharing

management, and multi-stage validation to create a robust, scalable digital health platform.

4.4.1 Problem Understanding

Healthcare systems face fragmented and incomplete patient data, which prevents smooth information sharing between providers, delays diagnosis, increases the risk of errors, and reduces the effectiveness of AI-driven decision support. Privacy issues, poor data sharing, limited AI testing, and unequal access, especially in rural areas, remain challenges. MedX's requirements, shown in Tables 3 and 4, were shaped through surveys, interviews, focus groups, and studies, ensuring usability, clinical relevance, and compatibility with healthcare systems. The dataset included user demographics, health records, and environmental data.

This section outlines the core algorithms implemented within the MedX platform for intelligent health assistance. These include disease prediction based on user symptoms, BMI-based diet recommendations, and real-time environmental health alerts.

Table 3: Core Application Capabilities.

No.	Capability	Description
1	Disease Prediction	Uses AI algorithms on user-provided data and symptoms to suggest possible health conditions.
2	Real-time Environmental Integration	Delivers live environmental data (temperature, AQI) to guide health-related decisions.
3	Secure Medical Record Management	Encrypts and stores medical histories for easy and safe retrieval.
4	Doctor Consultation	Connects users with the top five local doctors, filtered by specialty.
5	BMI-Based Diet Plans	Generates personalized meal plans based on Body Mass Index.
6	Curated Health Tips	Provides practical advice for wellness areas like skincare, dental care, and hair health.

Table 4: System Quality and Performance Criteria.

No.	Criterion	Description
1	High Accessibility	Works across various devices and offers a user-friendly interface for diverse demographics.
2	Scalability	Handles increasing user numbers and future feature integration without performance loss.
3	Security	Uses secure protocols (HTTPS, TLS) and authentication to protect sensitive health data.
4	Performance	Optimized for fast load times and minimal latency, even under heavy traffic.
5	Availability	Maintains high uptime with cloud-based infrastructure and ensures compliance with GDPR and HIPAA standards.

4.4.2 Disease Prediction Algorithm

The algorithm utilizes user-reported symptoms and leverages Infermedica's medical inference API to predict potential diseases.

Algorithm 1: Symptom-Based Disease Prediction

Input: A collection of user-reported symptoms S = {s1, s2, ..., sn}

Output: Predicted disease D

1. Start and establish a connection to the medical inference API using valid authentication credentials.
2. Transmit the symptom set S to the API for evaluation.
3. Obtain from the API a ranked list of potential diseases along with their associated confidence levels.
4. For each identified condition DI in the returned list:
   a. If confidence (di) is greater than or equal to the predefined threshold:
      i. Assign D←di
      ii. Stop further evaluation.
5. Return the predicted disease D.

4.5 BMI-Based Diet Plan Generator Using Random Forest

This model applies a supervised machine learning approach to recommend diet plans. A Random Forest classifier is trained on labeled health data, which includes BMI values and corresponding diet categories.

4.6 Environmental Risk Alert Algorithm Using Adversarial Generative Modeling

This algorithm leverages a conditional Generative Adversarial Network (cGAN) to model normal environmental patterns and detect anomalous or high-risk conditions for health. The generator (G) learns to produce realistic environmental time-series segments conditioned on context (e.g., location, season, hour), while the discriminator (D) learns to distinguish real from generated segments. At inference, anomaly and risk scores derived from D (and optional reconstruction error) are used to trigger health alerts with recommended actions. Figure 2 illustrates the workflow of Conditional Generative Adversarial Networks (cGAN) integrated with machine learning. Input data, along with conditions, are fed into the generator, which produces synthetic data. The discriminator evaluates the generator's output against real data to improve accuracy. The refined data is then processed by a machine learning model, ultimately generating reliable outputs. This workflow highlights the synergy of cGAN's data augmentation and machine learning's predictive power in delivering optimized solutions.

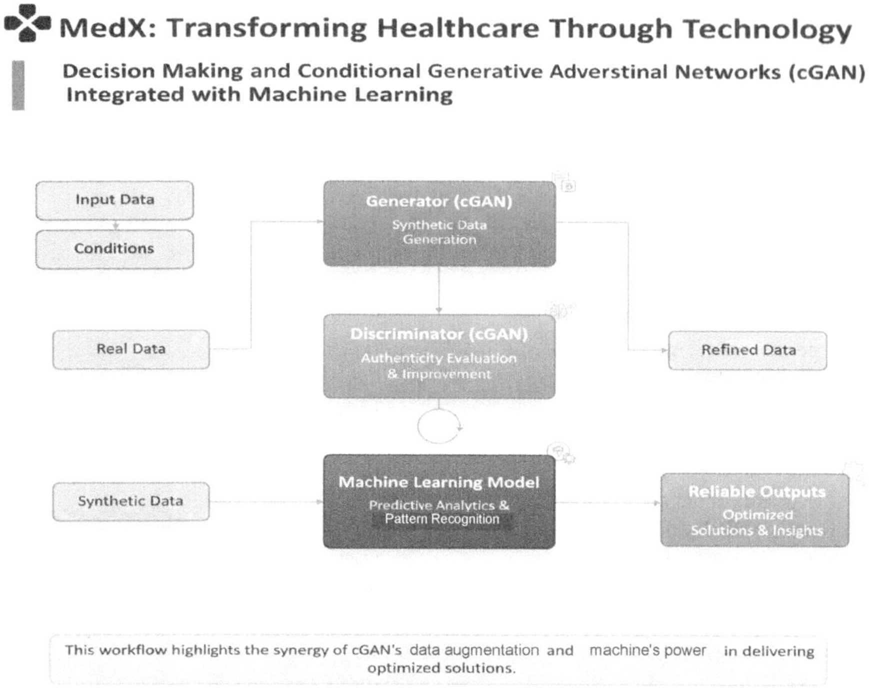

Figure 2: Workflow Using cGAN and Machine Learning.

Algorithm 2: Environmental Health Risk Assessment via GAN

Input:

- Historical environmental data X (e.g., temperature, AQI, humidity, PM2.5, PM10, ozone, wind).
- Context variables C (location, season, weekday/weekend, hour).
- Real-time measurements x_t at time t with context c_t.
- Pretrained cGAN = {Generator G, Discriminator D}; thresholds τ_{anom}, τ_{alert}.

Output:

- Health Alert H with severity level and recommended guidance.
  Training (offline):
  (1) Preprocess data: Clean, impute missing values, normalize features; create sliding windows of length L.
  (2) Build cGAN: G(z, C) → synthetic segment; D(segment, C) → probability(real).
  (3) Optimize $\min_G \max_D [E_{real} \log D(x, C) + E_z \log(1 - D(G(z, C), C))]$ with regularization (gradient penalty, spectral norm).

(4) (Optional) Add reconstruction head R to G for pseudo-reconstruction loss to stabilize training.

(5) Calibrate thresholds τ_anom and τ_alert on a validation set using ROC/PR analysis and desired sensitivity/specificity.

Inference (online):

(1) Ingest latest window x_t (length L) with context c_t; normalize using training stats.

(2) Compute discriminator score: s_D = 1 − D(x_t, c_t). Higher values suggest abnormality.

(3) (Optional) Compute reconstruction error: x̂_t = G(z*, c_t) with z* estimated by latent optimization; e_rec = ||x_t − x̂_t||.

(4) Aggregate anomaly score: a_t = α·s_D + (1−α)·normalize(e_rec), α ∈ [0,1].

(5) Map anomaly to health risk: r_t = w^T·[a_t, AQI_t, HeatIndex_t, ColdIndex_t, TrendFlags_t].

(6) Determine alert level: If r_t ≥ τ_alert ⇒ H = {SEVERE, message}; elif a_t ≥ τ_anom ⇒ H = {MODERATE, message}; else H = {NORMAL, guidance}.

(7) Compose guidance: For heat (e.g., T ≥ 40°C), recommend hydration/shade; for poor AQI (e.g., AQI ≥ 200), recommend masks/limit outdoor time; for cold (T < 5°C), advise layering and indoor stay.

(8) Log decision with explanations: Key contributors (e.g., high AQI, rising trend), uncertainty estimate from Monte Carlo sampling over z.

Where:

- Use context conditioning (C) to reduce false positives from seasonal/diurnal shifts.
- Recalibrate thresholds periodically as sensors or locations change (dataset shift).
- Provide human-readable rationales (e.g., SHAP on r_t components) for clinician/user trust.
- Fail-safe: if sensors are unavailable or values are extreme/out-of-range, fall back to rule-based alerts.

5. Discussion

MedX represents a significant step forward in addressing key challenges within the healthcare system. By integrating AI-driven analytics and user-friendly interfaces, it enhances accessibility, engagement, and proactive health management. Recent additions, such as emergency consultation access, BMI-based diet plans, and environmental health insights, further differentiate MedX as a comprehensive healthcare solution. However, the success of MedX depends on addressing barriers like provider resistance, data privacy concerns, and technical scalability. As Miller (2022) points out, "Data security and interoperability must be prioritized to build

Chapter 4

Explainable AI for Clinicians

Bridging the Trust Gap in Smart Healthcare

Nargis Khaliq,[1,2] *Swati Sah,*[1,2,*] *Aditya Dayal Tyagi*[1,2] and *Himani Tyagi*[1,2]

1. Introduction

Artificial Intelligence (AI) systems, especially those based on deep learning, have shown remarkable performance in various medical tasks—from image analysis to predictive analytics. However, a key barrier to clinical adoption is the "black-box" nature of many AI models. Clinicians are often uncertain how a complex model arrived at a recommendation or diagnosis, leading to a gap between algorithmic outputs and human understanding. Explainable AI (XAI) has emerged to bridge this gap by making AI's decision processes more transparent and interpretable. In essence, XAI comprises methods and tools that explain or clarify how AI models reach their conclusions, providing convincing evidence or reasoning behind a model's prediction (Wiegreffe 2019). The goal is to mimic aspects of human reasoning—much like a doctor forms a diagnosis and then articulates the reasoning based on symptoms and test results (Holzinger 2019). In the healthcare domain, explainability is not just a nicety but often a necessity. Decisions in medicine are high-stakes, and clinicians are trained to trust processes they understand. Without clarity on why an AI suggests a certain diagnosis or treatment, clinicians may question its value or hesitate to use it in practice (Ribeiro 2016). Moreover, regulatory frameworks and ethical guidelines increasingly demand transparency; for example, the European Union's ethics guidelines for trustworthy AI emphasise transparency and accountability as core requirements for AI in healthcare. Early initiatives like DARPA's XAI programme

[1] Department of Computer Science and Applications, Sharda University, Greater Noida, India.
[2] Sharda School of Computing Science and Engineering, Sharda University, Greater Noida, India.
Emails: nargisbhat91@gmail.com, https://orcid.org/0009-0003-4391-0661; adityadayaltyagai@gmail.com, https://orchid.org/0009-0005-0557-2918; htyagi.iitkgp@gmail.com, https://orcid.org/0009-0004-6020-9874
* Corresponding author: iswatisah19@gmail.com, https://orcid.org/0000-0002-0142-4660
DOI: 10.1201/9781003783855-5

(launched in 2015) highlighted the importance of developing AI systems that can explain their reasoning, with the ultimate aim of fostering appropriate trust from users including clinicians (Rieger 2018). Since then, the medical AI community has seen a surge in research on XAI techniques, reflecting its critical role in safety-critical environments such as healthcare (mdpi.com). At its core, XAI in healthcare serves two broad user groups: AI developers and clinical end-users (Turek 2018). For AI experts and data scientists, XAI tools help debug models and ensure they focus on medically relevant patterns. For clinicians and healthcare providers, XAI provides insights that make the model's suggestions understandable in clinical terms – thereby enabling the clinician to validate the AI's reasoning against their own medical knowledge (Reis 2021). This dual utility of XAI helps align AI behaviour with clinical expectations, a key step towards integrating AI into routine care. Ultimately, by making AI's inner workings more transparent, XAI aims to build an appropriate level of trust—confidence when warranted, but also healthy scepticism when the AI might be wrong (Malinverno 2023). In the following sections, we delve deeper into the challenges of clinician trust, the arsenal of XAI techniques available, and how explainability can be woven into smart healthcare systems to bridge the trust gap between clinicians and AI.

1.1 Motivation of the Study

The last few years show that artificial intelligence (AI) is revolutionising healthcare. Improved diagnostic accuracy, better-planned treatments, and intensified patient monitoring make artificial intelligence a reality in practice. Yet most AI models have been deep learning black boxes, largely without a clear definition. Moreover, such transparency issues cause reluctance by clinics to fully trust AI tools and have adoption in daily medical practice.

Clinicians need to trust—just not always. Their decisions can directly affect patient safety, so they need AI systems that not only deliver accurate predictions but also clearly explain the reasoning behind them. This is where Explainable AI (XAI) becomes vital. By making AI's decision-making process understandable, XAI can help bridge the trust gap, enabling clinicians to work confidently alongside intelligent systems. Our impetus for this research is to investigate how XAI can enhance transparency, accountability, and ethical decision-making and ultimately lead to a more effective collaboration between humans and machines in smart healthcare.

2. Challenges in Clinical Trust and AI Adaptation

Artificial Intelligence (AI) systems, especially those based on deep learning, have shown remarkable performance in various medical tasks—from image analysis to predictive analytics. However, a key barrier to clinical adoption is the "black-box" nature of many AI models. Clinicians are often uncertain how a complex model arrived at a recommendation or diagnosis, leading to a gap between algorithmic outputs and human understanding. Explainable AI (XAI) has emerged to bridge this gap by making AI's decision processes more transparent and interpretable. In essence, XAI comprises methods and tools that explain or clarify how AI models reach their conclusions, providing convincing evidence or reasoning behind a model's prediction. The goal is to mimic aspects of human reasoning—much like a doctor forms a diagnosis and then articulates the reasoning based on symptoms and test results. In the healthcare domain, explainability is not just a nicety but often a necessity. Decisions in medicine are high-stakes, and clinicians are trained to trust processes they understand. Without clarity on why an AI suggests a certain diagnosis or treatment, clinicians may question its value or hesitate to use it in practice. Moreover, regulatory frameworks and ethical guidelines increasingly demand transparency; for example, the European Union's ethics guidelines for trustworthy AI emphasise transparency and accountability as core requirements for AI in healthcare. Early initiatives like DARPA's XAI programme (launched in 2015) highlighted the importance of developing AI systems that can explain their reasoning, with the ultimate aim of fostering appropriate trust from users including clinicians. Since then, the medical AI community has seen a surge in research on XAI techniques, reflecting its critical role in safety-critical environments. At its core, XAI in healthcare serves two broad user groups: AI developers and clinical end-users. For AI experts and data scientists, XAI tools help debug models and ensure they focus on medically relevant patterns. For clinicians and healthcare providers, XAI provides insights that make the model's suggestions understandable in clinical terms—thereby enabling the clinician to validate the AI's reasoning against their own medical knowledge. This dual utility of XAI helps align AI behaviour with clinical expectations, a key step towards integrating AI into routine care. Ultimately, by making AI's inner workings more transparent, XAI aims to build an appropriate level of trust—confidence when warranted, but also healthy scepticism when the AI might be wrong. In the following sections, we delve deeper into the challenges of clinician trust, the arsenal of XAI techniques available, and how explainability can be woven into smart healthcare systems to bridge the trust gap between clinicians and AI.

To make AI models interpretable to clinicians, a variety of XAI techniques have been developed. These techniques can be broadly categorised by the form of explanation they provide (visual, textual, feature-based, example-based, etc.) and whether they apply to any model (model-agnostic) or are specific to certain model types (model-specific). Below we summarise prominent XAI methods, each with a brief description:

Saliency Maps and Grad-CAM: Saliency maps highlight parts of an input (e.g., regions of an image) that strongly influence the model's prediction. In deep neural networks for images, a common approach is Gradient-weighted Class Activation Mapping (Grad-CAM), which uses the gradients flowing into the final convolutional layer to produce a coarse heatmap of "important" pixels (Guidotti 2018). For example, in a chest X-ray classified as pneumonia, a Grad-CAM heatmap might light up the region of the lungs where the model detected infiltrates. Grad-CAM is popular for its intuitive visual explanations and was found to be the most frequently used XAI method in a review of XAI in cardiac imaging (Holzinger 2019). Variants like Grad-CAM++ refine this by accounting for multiple occurrences of features, yielding more detailed localisation. Saliency-based methods are valuable in domains like radiology and pathology where the anatomical location the model is examining is as important as what it predicts.

SHAP (Shapley Additive Explanations): SHAP is a model-agnostic, feature attribution method grounded in cooperative game theory (Tjoa and Guan 2020). It explains a prediction by computing the contribution of each feature—e.g., a lab test value or symptom—to the difference between the model's output for a given case and a baseline output (Alexander 2020). For instance, in a sepsis-risk model, SHAP might reveal that an elevated lactate level and low blood pressure were the top contributors to a high-risk prediction. SHAP is powerful in that it provides consistent and theoretically sound attributions, and it can be applied to any model by repeatedly evaluating the model with different combinations of features. However, it can be computationally intensive, since it requires evaluating many combinations of features to approximate each feature's marginal contribution (Ghassemi 2021). In healthcare, SHAP is popular for explaining predictions from complex models using tabular clinical data or even signals like ECGs (where it can highlight which portions of the ECG wave most influenced an arrhythmia classification) (EL Shawi and Al-Mallah 2022).

LIME (Local Interpretable Model-agnostic Explanations): LIME is a technique that generates an interpretable surrogate model to explain an individual prediction. The idea is to perturb the input around the instance of interest and see how the model's predictions change. By sampling, say, hundreds of variations of the original patient's data (e.g., masking or altering certain features) and observing the outputs, LIME fits a simple model—often a linear model for tabular data or a simplified linear combination of superpixels for images—that approximates the complex model's behaviour in the local neighbourhood of that instance. The result is a set of weights or a small decision rule that explains which features were most influential for that specific prediction. For example, LIME might explain an AI's diagnosis of diabetes for a particular patient by saying it was primarily due to high BMI and glucose level, as evidenced by the surrogate linear model's coefficients. LIME's strength is its flexibility and intuitiveness—it treats the original model as a black box and works for image, text, or tabular data. However, the explanations can be sensitive to the sampling procedure; different runs might yield slightly different explanations if the perturbations or weighting scheme change, as found in the current inventory. Thus, stability can be an issue, and LIME explanations need to be interpreted with some caution.

Counterfactual Explanations: A counterfactual explanation answers the question, "What minimal change to the input would alter the model's prediction?". If the patients condition were slightly different, the diagnosis would have changed: "If this patient's X-ray did not show nodule Y, the model would no longer predict lung cancer" or "Had the patient's cholesterol been lower by 20 points, the risk score would drop into the low-risk category." Counterfactuals are appealing because they provide actionable insight—they indicate how the input could be different to achieve a different outcome, relating closely to decision-making. Techniques to find counterfactuals often involve optimising in input space, under certain constraints to keep changes realistic, to flip the model's output. In healthcare, counterfactual explanations could potentially suggest alternative diagnoses or highlight borderline

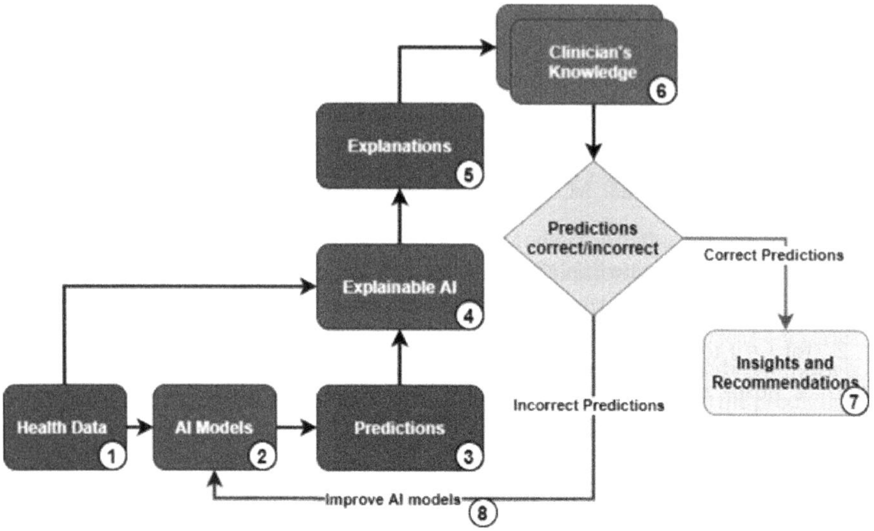

Figure 1: Illustrative workflow integrating XAI into clinical decision support.

explanation through a user interface, enabling them to interpret the AI's reasoning. The clinician can compare the AI's rationale with their own medical knowledge and patient context (Junaid 2023), which in turn informs their final decision (accepting, rejecting, or refining the AI's suggestion). By closing the loop in this manner, the system supports clinical decision-making in a way that is transparent and accountable. The successful integration of XAI into clinical decision support systems (CDSS) requires careful consideration of user experience, timing, and context. First and foremost, the explanations must be delivered in a clinically useful form. For example, in medical imaging AI applications, a common approach is to overlay the explanation (such as a heatmap) directly on the image being examined (datasciencecentral.com). Radiologists using an AI to detect tumours on MRI or nodules on X-rays will find a highlighted region much more intuitive than a raw probability score—the heatmap indicates where the model is looking and can be immediately checked against the radiologist's own visual assessment. This visual integration has the benefit of not significantly altering the workflow: the radiologist still reads the image, but now has an extra layer of insight. Similarly, for lab results or clinical predictions, integrating explanations means perhaps showing a ranked list of contributing factors right next to the model's output on the screen. If an AI alert flags a patient as high-risk for sepsis, the interface might display: Recent surgery, elevated lactate levels, and hypo tension emerged as the primary contributors to the prediction. This direct annotation of predictions with explanations helps clinicians understand why the alert is happening in real time, rather than having to trust a silent model output. Another crucial aspect is interactivity and customisation. Clinicians have varying levels of expertise with technology and different information needs. A well-designed XAI-enabled CDSS might allow the user to drill down for more information if needed. For instance, a physician might start with a simple explanation—"Feature X is high, contributing to diagnosis Y"—but could click to see more details or alternate explanation formats (perhaps switching from a feature list to a case-based explanation showing similar past patients). Providing such flexibility ensures that the explanation is neither too simplistic for expert users nor too overwhelming for novices. It also supports different reasoning approaches—some clinicians are numbers-oriented and might appreciate quantitative feature contributions, while others are visual or analogy-driven and might prefer exemplar cases or visual cues. Integration also involves validating the AI's reasoning within the clinical context. Explanations can sometimes reveal when an AI is focusing on spurious patterns. A famous example occurred in a pneumonia detection model for chest X-rays: the model appeared highly accurate, but explainability techniques showed it was relying on a hospital-specific artefact (a metal token in the corner of images) as a cheat, rather than actual medical features in the lungs (datasciencecentral.com, datasciencecentral.com). If such a model were integrated into a CDSS, a clinician seeing a heatmap lighting up the image corner (an irrelevant region) would immediately recognise the issue and distrust that prediction. This illustrates how XAI integration serves as a safety net—by exposing the model's attention and logic, it allows human experts to catch AI mistakes that would otherwise be hidden. In practice, when clinicians notice an explanation that "does not make sense" (e.g., highlighting an anatomically implausible region or citing a clinically insignificant variable), they can discount the AI's recommendation and potentially flag the model for review or improvement (Lundberg 2017). Thus, XAI not only builds trust but also helps in troubleshooting AI models and continuously improving them. Developers can gather feedback on explanations that failed to satisfy clinicians and refine the models to be more aligned with clinical reasoning. From a system architecture perspective, embedding XAI into CDSS means that for every prediction the AI makes, an explanation must be generated and delivered with low latency. Clinicians will not wait minutes for an explanation—it has to appear almost instantly with the result. This poses technical challenges since some explanation methods (like SHAP) are computationally heavy. Solutions include pre-computing explanations for common queries, using lighter approximations for real-time use, or deploying specialized explanation engines optimized for speed. Additionally, the system should log interactions with explanations (e.g., which parts of an explanation the clinician focused on) as this can inform future improvements and also serve documentation purposes (recording how AI advice was interpreted, which could be useful for auditing). In summary, integrating XAI into clinical workflows transforms AI from a mysterious oracle to a collaborative assistant. The AI provides not just answers but also reasoning, and the clinician remains in the loop, exercising judgment with the aid of machine intelligence. Early studies indicate that when explanations accompany AI recommendations, clinicians are more willing to consider and trust the AI's input, and are better at spotting when the AI might be wrong. This symbiosis—AI's computational power and speed combined with human expertise and oversight—is the essence of trustworthy smart healthcare. The next section will explore concrete case studies in various specialties, illustrating how XAI is being applied in practice.

4. Case Studies of XAI in Clinical Practice

4.1 Radiology: Explainable AI in Medical Imaging

Radiology has been at the forefront of medical AI, with numerous algorithms developed for detecting lesions, classifying diseases from scans, and even automating image segmentation. Given the visual nature of radiology, XAI techniques like saliency maps and Grad-CAM are especially prevalent. In practice: consider a deep learning model that analyses chest X-rays to detect pneumonia. If this model simply outputs "Pneumonia: Yes" with a probability, radiologists might be sceptical. But with XAI, the model can highlight the region in the lung fields where it found opacity consistent with pneumonia. This visual explanation allows the radiologist to verify that the AI is focusing on a plausible area (e.g., right lower lobe) rather than an artefact. It also provides a teaching point—perhaps drawing attention to a subtle finding the radiologist hadn't initially noted. Studies in cardiovascular imaging—which overlaps radiology and cardiology—show that such approaches are common: out of dozens of recent studies applying XAI to cardiac imaging (ultrasound, CT, MRI, X-ray), Grad-CAM was the most frequently employed method, followed by SHAP, and most studies produced saliency-based visual explanations for model outputs (Lundberg and Lee 2017). This indicates a heavy reliance on techniques that can generate heatmaps or attention maps on images, aligning with radiologists' need to see where the AI is "looking." One illustrative case comes from an algorithm developed to triage head CT scans for haemorrhages. The model would flag scans with potential bleeding. With explainability, each flagged scan came with a heatmap highlighting the region of suspected haemorrhage. Radiologists using the system reported that, even when the AI was wrong, the heatmap often pointed to an unusual feature that warranted a second look—occasionally catching a different pathology or a technical issue with the scan. This underscores that explanations can add value beyond the immediate task. On the flip side, as mentioned earlier, explainability also revealed a cautionary tale: the pneumonia detector that latched onto a confounding token marker. When deployed across hospitals, the model's performance dropped because it had learned "if token present, likely pneumonia" (an accidental correlation in the training set). A saliency map in that case essentially debugged the model by exposing its flawed reasoning, leading developers to retrain the model more carefully with that artifact removed. Overall, in radiology AI deployments, XAI is becoming standard. Vendors of AI radiology tools often include visualisation of the AI's focus areas as part of the product. Not only does this help in clinician acceptance, but it is increasingly demanded by regulators to ensure that automated image analysis is transparent and can be audited. Radiologists, being visual specialists, generally appreciate such explanations and have incorporated them as another image overlay to interpret. There remain challenges: sometimes saliency maps can be diffuse or hard to interpret, and they typically do not explain causation (they show association of regions with the prediction, not necessarily the specific features like "ground-glass opacity" or "calcification" that the model detected). Research is ongoing to develop more advanced XAI for imaging, such as methods that could generate textual explanations of an image finding ("the AI detected a mass in the left upper lobe with spiculated margins") or hybrid approaches that link visual features with clinical data.

4.2 Pathology: Explainable AI in Digital Pathology

Digital pathology involves analysing high-resolution images of tissue slides (histology) for tasks like cancer detection and grading. AI models in this field, often convolutional neural networks, can assist in identifying cancerous regions, classifying tumour subtypes, or counting mitotic figures. However, pathologists, like radiologists, are trained to trust what they see through the microscope. XAI has been applied to ensure AI highlights the relevant histological features. For instance, a deep learning model for prostate cancer Gleason grading might output a predicted Gleason score for each region of a biopsy. An explainable system will overlay colour-coded highlights on the digitised slide image, indicating which glandular structures or cell patterns led to a region being classified as Gleason grade 4 vs. grade 3, for example. In a recent study, pathologists were shown various XAI explanations (using methods such as Grad-CAM, Grad-CAM++, integrated gradients, etc.) on prostate biopsy images to evaluate whether it improved their trust in the AI's grading (Zhao 2020).

The results were promising: the majority of the 15 pathologists found that the AI's assistance with explanations was helpful (Figure 2) and the method Grad-CAM++ received the highest usability scores. The visual overlays (heatmaps) on the histology slides helped them see if the AI was focusing on the correct morphology (e.g., irregular glands for higher Gleason grades). The study concluded that explainability indeed improved acceptance of the AI among users, by demystifying the AI's decisions and aligning them with the pathologists' own reasoning process.

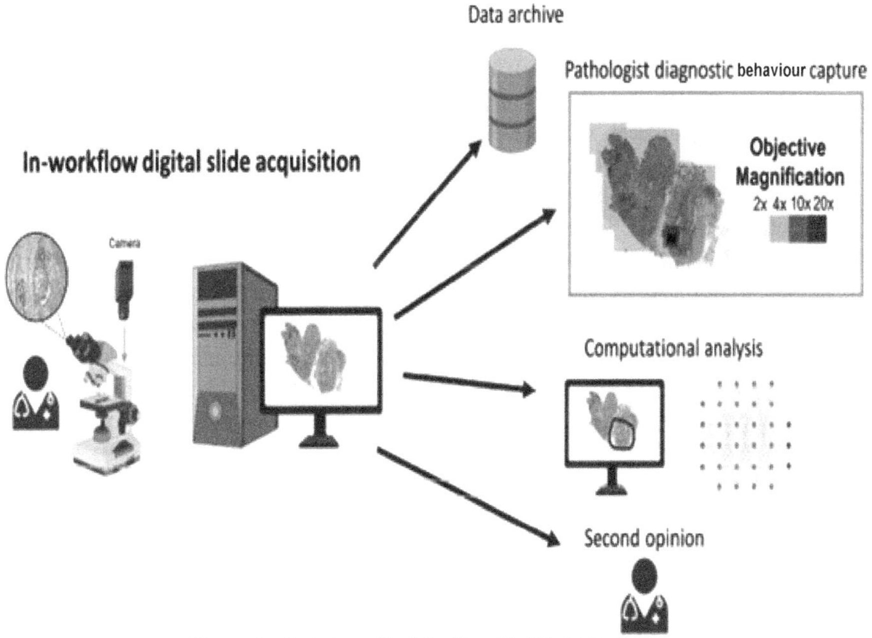

Figure 2: Pathology: Explainable AI in Digital Pathology.

cardiology, risk models are available to predict the five-year risk for an individual to develop any type of cardiac event. Such risk models are usually multi-variable models where patient demographics, laboratory values, imaging findings, etc., are included in the model's inputs. Explainable artificial intelligence (XAI) for these kinds of models is usually through model interpretation techniques such as feature attribution (SHAP, LIME) or rule extraction. Sometimes, these explanations portray surprising insight. For example, an explainable model of risk concerning potential heart attack may mention not only classical risk factors (blood pressure, cholesterol) but also a certain factor derived from EHR (for instance, many visits to cardiology before or taking a certain medication) as strong contributory—possibly uncovering proxy variables that the model is using. Clinicians would then investigate whether those proxies make sense or if they reflect some bias in data collection (e.g., more data on sicker patients leading the model to partly use "number of doctor visits" as a risk factor—which might be valid as a frailty indicator, or might be a bias). As a concrete example: consider an AI that processes cardiac MRI to detect heart muscle scarring (fibrosis). Suppose it highly confidently predicts about the myocardial fibrosis of a patient. It may then, for an XAI explanation, overlay a heatmap on the MRI showing the posterior wall of the left ventricle as the region where the model sees fibrosis, and it might additionally retrieve a couple of prior cases of fibrosis in that same location for comparison. The cardiologist, by having the explanation, can then verify against the late gadolinium enhancement images (the gold standard for fibrosis) whether the AI-highlighted region corresponds to an actual hyperenhancement. If it does, the cardiologist thinks that the AI is effectively mimicking expert reading; if it does not, the cardiologist knows to be sceptical about that particular result. In summary, XAI functions as the bridge between complex AI models and clinicians' trust, across radiology, pathology, and cardiology. Each of the fields emphasises differently how explanations should match with the modality: vision-oriented explanations for image-heavy specialties and feature/logic-oriented explanations for data-driven predictions (Tjoa and Guan 2020b). The case studies so far have portrayed an increased sense of satisfaction and insight on the part of the user where XAI is concerned. On the other hand, it raises some questions with regard to the need for a more systematic approach to assessing whether such explanations really affect clinical outcomes (not merely the satisfaction level) – do radiologists indeed diagnose better or faster if assisted by AI+XAI? These questions will emerge in the following sections on wider issues of regulation and ethics and evaluation frameworks for trustworthy AI in healthcare.

5. Regulatory and Ethical Considerations

A close interplay exists between AI deployment in healthcare and regulatory or ethical requirements, many of which have directly motivated the use of explainable AI. Regulators and professional bodies recognize that trustworthy AI should be transparent, accountable, and compliant with ethical principles of medicine (beneficence, non-maleficence, autonomy, and justice). Explainability supports these aims well.

Regulatory Drivers for XAI: In the European Union, having been interpreted by some, the General Data Protection Regulation could grant people "the right to an explanation" concerning algorithmic decisions, especially in sensitive areas such as health (Cai 2019). More specifically within healthcare, there is a requirement in the EU for AI-based medical device regulation (i.e., Medical Device Regulation) to ensure that any information generated to assist clinicians by providing decision-support recommendations is sufficient information for that user to interpret and to use it safely. This effectively mandates a degree of transparency—a black-box system that outputs decisions without rationale would not satisfy this requirement. In the United States, the FDA has issued guiding principles for AI/ML in medical devices which emphasize transparency, the ability to interpret predictions, and the importance of algorithmic accountability throughout the product lifecycle. The FDA has not gone so far as to explicitly require XAI, but in practice, for an AI system to gain clinician and FDA acceptance, demonstrating interpretability and the capacity for validation by medical experts is a significant advantage. This has led many AI developers seeking approval to include interpretable models or explanation interfaces as part of their submissions.

Ethical Use and "Do No Harm": One of the core ethical principles in medicine is *primum non nocere*—first, do no harm. A poorly understood AI system can inadvertently cause harm (for example, by issuing erroneous recommendations that clinicians might follow). Explainability enhances safety by allowing improper model behaviours to be detected and corrected before they impact patient care. For example, if an ICU monitoring AI begins issuing excessive alerts, clear explanations help determine whether this reflects genuine patient deterioration or a technical issue such as sensor malfunction. Ethically, clinicians are also obligated to obtain informed consent. If an AI recommends a treatment change, the clinician must be able to explain why. Saying "the computer recommends this, but I do not know why" is ethically unacceptable. XAI therefore indirectly supports patient autonomy and informed decision-making.

Bias, Fairness, and Transparency: Another ethical dimension is fairness—AI models can inadvertently perpetuate or amplify biases present in training data, leading to disparate performance across different patient groups. Explainability can help audit models for such biases. By examining explanations for different subgroups, developers and clinicians might notice if a model is relying on a feature that is a proxy for race or gender in a problematic way (Saijo 2021). For example, if explanations for many African-American patients highlight a particular lab test influenced by socioeconomic factors, this may indicate a confounded pattern rather than a genuine clinical association. Some XAI methods are specifically geared towards global understanding of a model (like decision rules extraction), which can be used to present the model's logic in a transparent form for evaluation against ethical standards (e.g., checking if any rule differentiates on sensitive attributes without justification). At the same time, it must be acknowledged that explainability alone does not guarantee ethical AI. There is a risk of false reassurance—an explanation may appear plausible even if the model itself is making a flawed decision. As noted in a recent perspective, explanations can sometimes mislead. For example, an AI might provide an explanation that sounds reasonable, but is actually not the true cause of its decision (this can happen if the XAI method is imperfect). If clinicians place blind trust in the explanation without further scrutiny, that could be dangerous.

Therefore, a cautious approach is needed: XAI should be seen as a tool to assist human judgment, not replace it. From a regulatory standpoint, there are calls for standards to evaluate the quality of explanations themselves, to ensure they are not just present, but also accurate and useful.

Accountability and Legal Liability: Explainability contributes to clearer accountability. If an AI provides an explanation, it becomes easier to pinpoint whether an error was due to a model flaw, misuse by the clinician, or an unforeseeable scenario. This clarity is important for learning from mistakes and for legal processes. The presence of explanations might eventually influence malpractice considerations—for instance, if a clinician overrode an AI recommendation, a documented explanation could support why they did so (perhaps the explanation revealed a nuance the clinician judged as irrelevant or incorrect). Conversely, if a clinician followed an AI advice, the explanation record shows they did due diligence in reviewing the rationale (Saijo 2021). These are evolving areas, and legal frameworks are still catching up with AI technology.

Privacy and Data Governance: One ethical concern with XAI is that some techniques could potentially expose patient data. For example, case-based explanations that retrieve similar patient cases must be handled in compliance with privacy regulations (you wouldn't want to show identifiable details of a different patient as part of an explanation). Similarly, if a model explanation reveals a certain rare combination of features leading to a diagnosis, could that be identifying a patient cohort? Generally, the focus is to ensure that while AI models remain privacy-preserving, their explanations do not inadvertently leak sensitive information. This is a manageable issue—de-identified or synthesised examples can be used for case-based explanations, and feature-based explanations can be presented at an aggregate level when needed. In conclusion, the regulatory and ethical landscape is strongly motivating the adoption of XAI in healthcare AI systems. The consensus is that AI tools must be transparent, explainable, and under human control to be trustworthy. Bodies like the EU High-Level Expert Group on AI have put forward principles like the seven key requirements for trustworthy AI (human oversight, technical robustness, privacy, transparency, fairness, societal well-being, accountability), and explainability is a thread running through many of these. Future regulations (such as the EU's proposed AI Act) are likely to cement these into law for high-risk AI systems like those in medicine. Ethically, explainability aligns AI with the values of medical practice, ensuring that technology serves as a partner to clinicians and patients rather than an opaque authority. The onus is now on developers and the research community to develop XAI methods that are not only technically sound but also practically meet these regulatory and ethical expectations.

6. Frameworks for Evaluating Trust, Transparency, and Usability

As XAI techniques proliferate, a critical question arises: How do we evaluate the effectiveness of these explanations, especially in terms of trust, transparency, and real-world usability? Simply adding an explanation module to an AI system does not automatically guarantee that clinicians will trust or even properly understand the system. Thus, researchers have begun developing frameworks and metrics to systematically assess explainable AI in healthcare settings. One recent effort is the development of a clinician-informed evaluation checklist called CLIX-M (Clinician-informed XAI evaluation checklist with Metrics). This checklist, published in 2025, consists of 14 items covering clinical relevance, technical transparency, and decision-making impact. The idea is to have a standardised way to report and evaluate explanations: for example, items might include whether the explanation is consistent, whether it is interpretable to the target user (a junior doctor vs. a specialist might have different needs), and whether the explanation actually improved the decision outcome in testing. The CLIX-M checklist is a step toward XAI standardisation, aiming to enhance trust and reduce risks by ensuring explanations are of high quality. While it doesn't yet provide a single metric, it encourages thorough documentation and assessment across multiple dimensions of explainability. Another facet of evaluation is quantitative metrics for explanations. Researchers like Nauta et al. have outlined properties such as faithfulness, accuracy, stability, consistency, etc., that an explanation can be evaluated on in a laboratory setting. Faithfulness refers to how accurately the explanation reflects what the model actually used in its internal reasoning—e.g., if a model truly uses Feature A to make a decision, a faithful explanation should assign appropriate importance to Feature A. Stability or consistency might measure if small changes in input (or equivalent cases) lead to wildly different explanations, which would be undesirable. Some metrics exist: for instance, one can remove the top factors identified by an explanation and see how much the model output changes (a big change would indicate those factors were indeed important, supporting the explanation's correctness). There are also sanity checks for visual explanations – like randomised testing to ensure the heatmaps aren't just random or correlated with the original image regardless of model. However, purely quantitative evaluation misses the human aspect (Nohara et al. 2022). Therefore, human-centred evaluation is key. This involves user studies where clinicians interact with AI systems with and without explanations. Measures might include: Does the explanation improve the clinician's accuracy in using the AI (e.g., do they catch AI errors more often)? Does it improve efficiency (e.g., they make decisions faster or with less mental effort)? Are clinicians satisfied with the explanations, and do they find them useful or trustworthy? Tools like the System Usability Scale (SUS) or custom trust questionnaires are used. For trust, some studies measure cognition-based trust (confidence in the system's competence) and affect-based trust (a more emotional comfort level with the system). In the systematic review on XAI and clinician trust discussed earlier, the authors noted that most studies used cognitive trust measures (like rating how much they agree the AI is reliable) and very few addressed affective trust. Going forward, a robust evaluation framework would likely incorporate both: objective performance metrics (did XAI lead to better decisions?) and subjective ratings (do clinicians feel more confident and comfortable with the AI?). A challenge in creating universal frameworks is that trust and explainability are context-dependent. What fosters trust in one scenario (say, an AI that reads images for an ophthalmologist) might differ in another (an AI recommending chemotherapy doses to an oncologist). Thus, frameworks often emphasize involving domain experts in the evaluation design—as was done in CLIX-M by including clinicians to identify what they care about in an explanation. Moreover, trust is not binary, and an evaluation framework should

7.2 Personalization of Explanations

Not all users need the same type of explanation. Future XAI in medicine will likely adapt to the user's background. A medical student might get a more detailed, tutorial-like explanation from an AI (including basic definitions or anatomical reminders), whereas an expert surgeon might receive a terse explanation focusing on novel insights the model found. Similarly, patients themselves might be consumers of AI explanations—for instance, if an AI is used in direct-to-patient settings like symptom checkers or personalised risk reports. In such cases, explanations need to be translated into layperson-friendly language. Developing XAI that can toggle between different vocabularies and levels of detail (without losing fidelity) is an important area. This touches on natural language generation for explanations and perhaps integrating with medical knowledge graphs to generate richer rationales ("The AI noted your age and blood pressure, comparing it to patterns seen in thousands of similar patients, which is why it recommends you start hypertension medication").

7.3 Beyond Saliency—Causal and Counterfactual Explanations

Many current explanation techniques are associative—they show correlations or feature importance. The future will see a push towards causal explanations. This involves AI that can reason about cause and effect, not just correlation. For example, instead of a heatmap that highlights an area of an image, a causal explanation system might be able to say "The presence of this shadow caused the model to increase the probability of disease X, because in the training data that shadow was often due to artefact Y." Achieving true causal understanding in AI is extremely challenging, but important for trust—clinicians think in terms of causes (pathophysiology) not just correlations. Counterfactual explanations, as previously discussed, are one step in this direction, and we anticipate more sophisticated generation of what-if scenarios to answer clinicians' hypotheticals (e.g., "If the patient were 10 years younger with the same symptoms, would the diagnosis likely be different?").

7.4 Validation through Clinical Trials

While many studies have shown improved trust or satisfaction with XAI, the ultimate future direction is to validate in prospective clinical trials that XAI actually improves patient outcomes or process efficiency. For example, a trial could randomise some clinics to use an AI diagnostic tool with explanations and others without, and then measure differences in diagnostic accuracy, speed, or patient throughput. It could also measure adverse events or errors. Only through such evidence will we truly know how much explainability is needed and in what form. Perhaps certain routine tasks are fine with a mostly automated system (freeing up clinician time), whereas critical decisions absolutely require full transparency. The evidence base will guide these practices and also inform regulators on what to mandate. As one review concluded, future research should focus on robust, quantitative assessment of explainability and prospective clinical validation in real workflows.

7.5 Advanced XAI Techniques and Beyond-Accuracy Metrics

On the technical front, we expect new XAI techniques tailored to complex data like genomics, multi-modal data (combining imaging, text, waveform), and time-series (like ICU vital sign trends). Trustworthy AI will also involve uncertainty

Table 2: Advanced XAI Techniques and Beyond-Accuracy Evaluation Metrics.

Category	Technique/Metric	Description	Common Applications/ Examples
Model-Agnostic XAI	**SHAP (Shapley Additive Explanations)**	Feature attribution method based on game theory	Risk models, tabular EMR data, ECGs
	LIME (Local Interpretable Model-agnostic Explanations)	Perturbs inputs to learn local linear model around a prediction	Diagnosis explanation (e.g., diabetes, cancer)
	Anchors	High-precision rules that "anchor" predictions	High-confidence decision regions
Model-Specific XAI	**Grad-CAM/Grad-CAM++**	Highlights relevant areas in medical images using gradients	Radiology, pathology, dermatology
	Integrated Gradients	Cumulative gradient method for feature attribution	Deep learning on imaging, EHR data
	Prototype Learning/ Prototypical Parts	Learns interpretable prototypes in image space	Histopathology, dermatology
Beyond-Accuracy Metrics	**Fidelity**	How well the explanation reflects the model's decision-making	Used to validate SHAP, LIME, etc.
	Stability/Robustness	Explanation consistency across input perturbations	Perturbation tests on medical input data
	Infidelity/Sensitivity	Measures mismatch between explanation and true influence of input features	Saliency methods evaluation
	Simulatability	Examines whether users can approximate the model's prediction by applying the information provided in the explanation.	Clinical training simulations
	Trust Calibration	Confidence alignment between model, explanation, and user trust	Human-AI collaborative diagnosis
	Comprehensibility	How understandable the explanation is to domain users (clinicians)	Measured via surveys, usability scales
Regulatory Alignment	**Actionability/Decision Support**	Does explanation lead to clinically meaningful action?	FDA, EU compliance audits, usability testing

quantification—conveying how confident the model is in its prediction. An AI that can say "I'm not sure" and explain why (maybe the input data is out of distribution or missing some key pieces) will be far more trustworthy than one that always makes a guess. So, we might see "explainable uncertainty" visualisations that highlight missing information or conflicting evidence in a case mentioned in Table 2. Moreover, as AI moves into areas like robotics (surgical robots, AI-assisted procedures), explainability could involve showing the predicted outcome of different action plans (like a robot surgeon highlighting the anatomical structures it plans to avoid and why). In terms of metrics, the community is likely to establish standard benchmarks for explainability. Just as accuracy is benchmarked on test datasets, explanations might be benchmarked on how well they align with expert explanations. There could be competitions where models are evaluated not just on accuracy but also on the quality of their explanations using standardised tests, e.g., can the AI's explanation predict what a panel of doctors would say about the case?

7.6 Trust Calibration and Ethical AI Culture

Finally, the future of trustworthy AI in medicine involves a cultural shift. Clinicians will need training in AI literacy—including understanding the limitations of XAI (for instance, knowing that a saliency map is not infallible, or that SHAP values have uncertainty). Medical education might incorporate AI decision support and interpretation as part of the curriculum. On the flip side, data scientists will benefit from more exposure to clinical reasoning so they can design explanations that make sense to clinicians (Chen et al. 2022). Hospital systems might employ "clinical AI officers" whose job is to oversee AI tool deployment, monitor their performance, collect user feedback, and ensure that trust is being maintained appropriately (neither blind trust nor undue mistrust). In the near future, we might see guidelines from medical professional societies on the use of AI and XAI—for example, radiology boards recommending that any AI output used clinically should be reviewable with a heatmap explanation, or internal medicine boards suggesting that risk scores come with a feature contribution breakdown. In summary, the future of explainable, trustworthy AI in healthcare is one where AI is seamlessly embedded in clinical practice as a transparent collaborator. Achieving this means advancing our technology (for better explanations), our evaluation methods (to ensure those explanations truly help), and our human systems (training and processes to utilise AI wisely). If we succeed, we stand to unlock the full potential of smart healthcare—AI augmenting clinicians in delivering faster, safer, and more personalised care—without sacrificing the human trust at the core of the patient-clinician relationship.

8. Conclusion

Explainable AI in healthcare is transforming the interaction between clinicians and intelligent systems from one of scepticism and mystery to one of cautious trust and synergy. In this chapter, we discussed how XAI techniques (like SHAP, LIME, saliency maps, and counterfactuals) provide the much-needed transparency to AI's decision-making process, enabling clinicians to understand why a model is suggesting a particular diagnosis or treatment. We explored the multifaceted challenge of clinician trust—emphasizing that the goal is calibrated trust, not uncritical acceptance. Integration of XAI into clinical workflows is critical: explanations must be delivered in intuitive ways—visual overlays, natural language, etc.—at the right time to support decision-making without overload. Case studies in radiology, pathology, and cardiology illustrate both the power and current limitations of XAI in real-world applications. On the regulatory and ethical front, we see a clear mandate for transparency and the development of standards to ensure AI systems do no harm and can be held accountable. Frameworks for evaluating explainability are emerging, blending technical metrics with human factors assessment, to quantitatively and qualitatively ensure that XAI methods are actually effective in building trust and improving decisions. Finally, we outlined future directions—pointing to more interactive, personalised explanations, the need for rigorous clinical validation of XAI, and a vision of truly trustworthy AI that complements human clinicians. Bridging the trust gap is an ongoing journey, but with explainable AI serving as the bridge, the destination of widely accepted and ethically sound AI in healthcare comes into clearer view. By continuing to refine XAI techniques and grounding them in the realities of clinical practice, we move closer to AI systems that clinicians not only use, but genuinely trust as partners in improving patient care.

References

Alexander, G. L., Powell, K. R., and Deroche, C. B. (2020). An evaluation of telehealth expansion in U.S. nursing homes. *Journal of the American Medical Informatics Association*, 28(2). doi: https://doi.org/10.1093/jamia/ocaa253.

Cai, C. J., Winter, S., Steiner, D., Wilcox, L., and Terry, M. (2019). 'Hello AI': Uncovering the Onboarding Needs of Medical Practitioners for Human-AI Collaborative Decision-Making. Proceedings of the ACM on Human-Computer Interaction, 3(CSCW), pp. 1–24. doi: https://doi.org/10.1145/3359206.

Chen, C., Li, O., Tao, C., Alina Jade Barnett, Su, J., and Rudin, C. (2018). This looks like that: Deep learning for interpretable image recognition. *arXiv (Cornell University)*. doi: https://doi.org/10.48550/arxiv.1806.10574.

Chen, C., Li, O., Tao, D., Barnett, A., Rudin, C., and Su, J. K. (2022). This looks like that: Deep learning for interpretable image recognition. *Nature Machine Intelligence*, 4: 55–67. https://doi.org/10.1038/s42256-021-00455-4

Chen, J., and Liu, J. (2022). Management of nonmotor symptoms in parkinson disease. *Journal of Innovations in Medical Research*, 1(5): 18–33. doi: https://doi.org/10.56397/jimr/2022.12.03.

Current Research Inventory. (1977). Latin American Research Review, 12(3), pp. 215–285. doi: https://doi.org/10.1017/s0023879100030612.

EL Shawi, R., and Al-Mallah, M. H. (2022). Interpretable local concept-based explanation with human feedback to predict all-cause mortality. *Journal of Artificial Intelligence Research*, 75: 833–855. doi: https://doi.org/10.1613/jair.1.14019.

Ghassemi, M., Oakden-Rayner, L., and Beam, A. L. (2021). The false hope of current approaches to explainable artificial intelligence in health care. *The Lancet Digital Health*, [online] 3(11): e745–e750. doi: https://doi.org/10.1016/S2589-7500(21)00208-9.

Gilpin, L. H., Bau, D., Yuan, B. Z., Bajwa, A., Specter, M., and Kagal, L. (2018). Explaining explanations: An overview of interpretability of machine learning. *2018 IEEE 5th International Conference on Data Science and Advanced Analytics* (DSAA). doi: https://doi.org/10.1109/dsaa.2018.00018.

Guidotti, R., Monreale, A., Ruggieri, S., Turini, F., Giannotti, F., and Pedreschi, D. (2018). A Survey of Methods for Explaining Black Box Models. *ACM Computing Surveys*, 51(5): 1–42. doi: https://doi.org/10.1145/3236009.

Section 2
Methods and Principles

Chapter 5

Heart Disease Prediction Using Machine Learning

Shachi Mall,[1,*] *Ravindra Nath*[2] and *Prashant Srivastava*[3]

1. Introduction

Cardiovascular diseases (CVDs) are recognized as a major cause of mortality worldwide. WHO reports indicate that around 17.9 million deaths occur annually due to these conditions, representing roughly one-third of all deaths across the globe (WHO 2021). Importantly, CVD is not a singular condition; rather, it is a broad term encompassing various CVD disorders. Given the significant global mortality burden and its rising trend, the timely identification of cardiac failure is critical. Timely identification enables medical interventions that can improve patient outcomes and reduce fatality rates. To identify CVD is a complex task, and numerous studies have sought to identify the most relevant clinical features associated with the condition (Chicco and Jurman 2020). Computational learning methods, such as advanced neural architectures, have demonstrated strong potential in forecasting CVD (Ghosh et al. 2021). While earlier studies have explored a variety of ML algorithms, many relied on basic or conventional models for comparison (Chen et al. 2020). This gap has motivated the present study to examine state-of-the-art machine learning techniques and assess their performance in predicting heart failure (Chen et al. 2020). The primary objective of this work is to identify the best-performing machine learning algorithms using a well-known benchmark dataset (Litjens et al. 2017). By systematically applying and evaluating a range of ML models, we aim to develop a robust predictive framework for heart failure risk assessment (Ahmad et al. 2020). The growing global burden of CVDs makes early prediction and diagnosis a critical

[1] Department of Computer Science and Engineering, Galgotias University Greater Noida, India.
[2] Department of Computer Science and Engineering, BBA University Lucknow, India.
Email: rnkatiyar@gmail.com, https://orcid.org/.0009-0008-2811-3716
[3] Department of Computer Science and Engineering, CSJM University Kanpur, India.
Email: theprashant1@gmail.com, https://orcid.org/0000-0002-6607-530X
* Corresponding author: shachimall@gmail.com, https://orcid.org/0000-0002-4443-4885

DOI: 10.1201/9781003783855-7

Algorithm: Decision Tree Classifier

Input: Training data (X, y), splitting criterion (e.g., Gini, Entropy)

Output: Decision tree model

1. If all samples belong to the same class, create a leaf node with that class
2. Else:
   a. Select the best feature to split based on the chosen criterion
   b. Partition the dataset into subsets based on feature values
   c. Recursively apply the algorithm to each subset
3. Stop splitting when:
   - Maximum depth is reached, or
   - Node has too few samples, or
   - No improvement in split criterion
4. Assign leaf nodes with the most frequent class label in the subset

Artificial Neural Network (ANN): A network of interconnected computational units (neurons) organized in layers for learning complex relationships in data. Used for classification and regression tasks.

Algorithm: Artificial Neural Network Classifier

Input: Training data (X, y), number of hidden layers L, learning rate α, epochs E

Output: Trained network parameters (weights, biases)

1. Initialize weights and biases for all layers
2. For epoch = 1 to E
   a. Forward Pass:
      - For each layer l in 1 to L:
        i. Compute weighted sum: z_l = w_l · a_(l-1) + b_l
        ii. Apply activation function to get a_l
   b. Compute output ŷ and calculate loss
   c. Backward Pass:
      - Compute gradients using backpropagation
      - Update weights and biases using gradient descent
3. Return trained weights and biases

5. Methodology

The experimental process was carried out in the following steps:

- **Initial Review:** The dataset was examined manually to understand its structure and to interpret the meaning of each feature.
- **Data Preprocessing:** The data was cleaned to handle missing entries, incorrect values, mismatches with the feature descriptions, and outliers. Python's *pandas* library was used to perform these operations.

- **Data Splitting:** The dataset was divided into two parts - 80% for training and 20% for testing. The split was done randomly to ensure unbiased evaluation.
- **Model Training and Evaluation:** Machine learning models were trained using the training set. Their performance was then assessed on the test set using appropriate evaluation metrics.

5.1 Dataset

We use our collected data for the prediction of the patient's health related to heart diseases. In this, we downloaded our dataset from www.kaggle.com website. This website provides a table of the records of a patient and the table contains many attributes such as (Kaggle n.d.) -:

1. Age of the individual
2. Biological sex
3. Type of chest pain (4 possible categories)
4. Resting blood pressure level
5. Serum cholesterol measured in mg/dl
6. Fasting blood sugar greater than 120 mg/dl (yes/no)
7. Resting electrocardiogram results (coded as 0, 1, or 2)
8. Maximum heart rate recorded during exercise
9. Presence of exercise-induced angina
10. "Oldpeak" – ST depression caused by exercise compared to rest
11. Slope of the peak exercise ST segment
12. Count of major blood vessels (0–3) highlighted by fluoroscopy
13. Thalassemia indicator: 3 = normal, 6 = fixed defect, 7 = reversible defect.

In this, we evaluate our model based on its performance, which can further help to determine the model's accuracy, reliability, and quality, and whether it's suitable for real-world applications.

- **Accuracy Score:** We have imported this accuracy from the sklearn.metrics library. As our model is asked to predict the target, the predicted values will be compared with the original target values. For example, if the total data is 100 and we are predicting 95 percent of values correctly, then our accuracy score is 95 percent. So the accuracy score is the number of correct values that we predict.
- **Building the Predictive System:** In this section, if we give the values of all the feature values such as the age of the person, their sex, the chest pain value, and all the attribute values, these should help us to predict the target values—that is, whether the person is suffering from heart disease or not.
- **Splitting the Data:** *In this section*, we split our data into two parts, which are training and testing, because we often train our machine learning algorithm with training data and we evaluate the performance of our model using the testing data, as shown in Figure 1, 2, and 3. This part is called train and test split because we split our data into two parts: training and testing.

```
#print first 5 rows of the dataset
data.head()
```

	age	sex	cp	trestbps	chol	fbs	restecg	thalach	exang	oldpeak	slope	ca	thal	target
0	63	1	3	145	233	1	0	150	0	2.3	0	0	1	1
1	37	1	2	130	250	0	1	187	0	3.5	0	0	2	1
2	41	0	1	130	204	0	0	172	0	1.4	2	0	2	1
3	56	1	1	120	236	0	1	178	0	0.8	2	0	2	1
4	57	0	0	120	354	0	1	163	1	0.6	2	0	2	1

Figure 1: This figure table indicates first five rows of the dataset.

```
# print last 5 rows of the dataset
data.tail()
```

	age	sex	cp	trestbps	chol	fbs	restecg	thalach	exang	oldpeak	slope	ca	thal	target
298	57	0	0	140	241	0	1	123	1	0.2	1	0	3	0
299	45	1	3	110	264	0	1	132	0	1.2	1	0	3	0
300	68	1	0	144	193	1	1	141	0	3.4	1	2	3	0
301	57	1	0	130	131	0	1	115	1	1.2	1	1	3	0
302	57	0	1	130	236	0	0	174	0	0.0	1	1	2	0

Figure 2: The second figure includes last five rows of the dataset.

```
# getting some info about the data
data.info()
```

```
<class 'pandas.core.frame.DataFrame'>
RangeIndex: 303 entries, 0 to 302
Data columns (total 14 columns):
 #   Column    Non-Null Count  Dtype
---  ------    --------------  -----
 0   age       303 non-null    int64
 1   sex       303 non-null    int64
 2   cp        303 non-null    int64
 3   trestbps  303 non-null    int64
 4   chol      303 non-null    int64
 5   fbs       303 non-null    int64
 6   restecg   303 non-null    int64
 7   thalach   303 non-null    int64
 8   exang     303 non-null    int64
 9   oldpeak   303 non-null    float64
 10  slope     303 non-null    int64
 11  ca        303 non-null    int64
 12  thal      303 non-null    int64
 13  target    303 non-null    int64
dtypes: float64(1), int64(13)
memory usage: 33.3 KB
```

Figure 3: Figure includes features of the dataset.

- **Data Collection and Processing:** In this section, we uploaded the data for the prediction that we previously downloaded from the Kaggle website. In this, we loaded our CSV file to a pandas dataframe. Following are the screenshots of the tables:

6. Results

In this study, we develop a machine learning-based model to accurately identify patients who are at risk of heart disease. The proposed framework is designed to serve as a reliable decision-support tool for clinicians, enabling them to take timely and appropriate medical actions based on the model's predictions. Our team focuses on building a Heart Disease Prediction System using machine learning techniques implemented in Python. The system is designed to collect relevant clinical data from patients and predict whether a heart disease condition is present, as shown in Figure 4. The binary classification approach is appropriate for this scenario, as it provides a clear yes/no result, aiding clinicians in making faster decisions. The model is developed and executed using Python in Google Colaboratory, an online cloud-based environment that supports large-scale data processing and machine learning experimentation. For the dataset, we utilize a publicly available source from Kaggle, comprising approximately 10,000 patient records. Each record includes multiple clinical attributes that are processed and analyzed to train the machine learning model. The specific algorithms, preprocessing steps, and evaluation metrics used in this system are discussed in later sections.

	age	sex	cp	trestbps	chol	fbs	restecg	thalach	exang
count	297.000000	297.000000	297.000000	297.000000	297.000000	297.000000	297.000000	297.000000	297.000000
mean	54.542088	0.676768	3.158249	131.693603	247.350168	0.144781	0.996633	149.599327	0.326599
std	9.049736	0.468500	0.964859	17.762806	51.997583	0.352474	0.994914	22.941562	0.469761
min	29.000000	0.000000	1.000000	94.000000	126.000000	0.000000	0.000000	71.000000	0.000000
25%	48.000000	0.000000	3.000000	120.000000	211.000000	0.000000	0.000000	133.000000	0.000000
50%	56.000000	1.000000	3.000000	130.000000	243.000000	0.000000	1.000000	153.000000	0.000000
75%	61.000000	1.000000	4.000000	140.000000	276.000000	0.000000	2.000000	166.000000	1.000000
max	77.000000	1.000000	4.000000	200.000000	564.000000	1.000000	2.000000	202.000000	1.000000

Figure 4: This table is used for the statistical measures about the data.

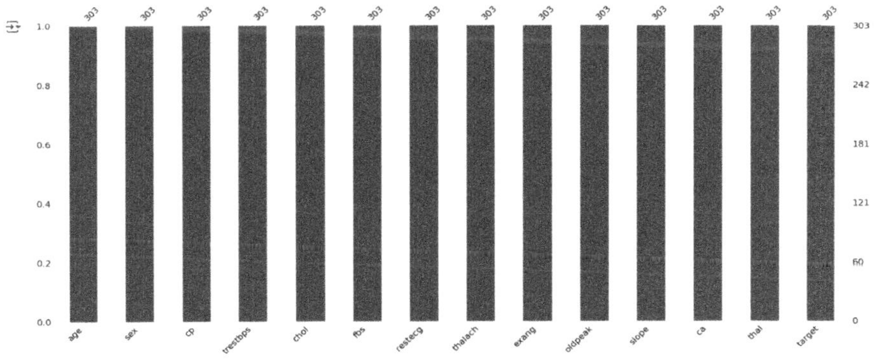

Graph 1: This graph is used for detecting the missing values.

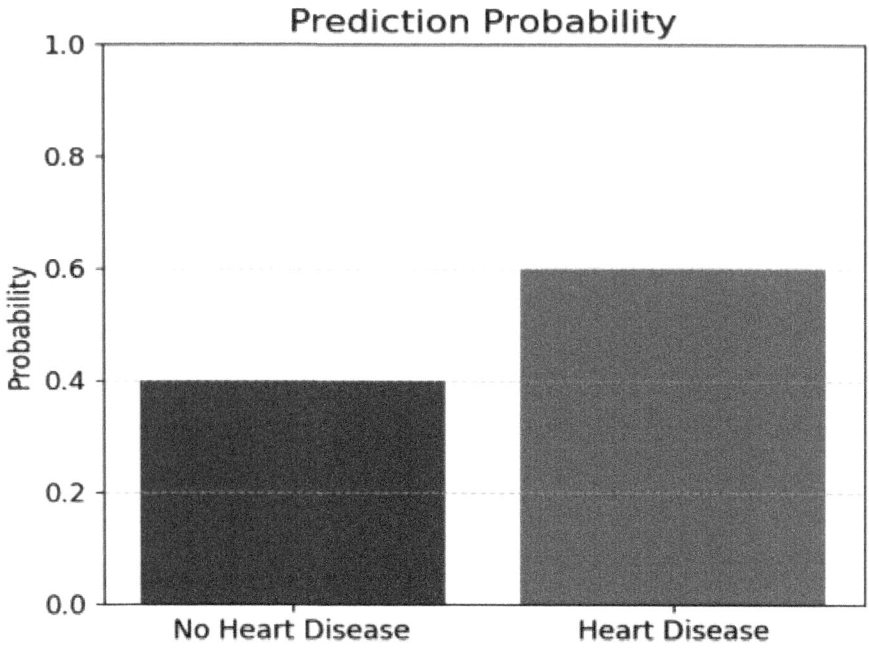

Graph 4: Visualization of the distribution in target variable barchart.

7. Conclusion

Early prediction and timely diagnosis of CVDs play a crucial role in reducing the number of fatalities associated with these conditions. In response to this challenge, numerous computational approaches have been proposed in the literature, targeting different aspects of predicting, detecting, and managing heart-related diseases. Much of the existing research has focused on identifying key features that influence the survival rate of patients with cardiovascular conditions. Figure 4 presents the statistical measures of the dataset, providing an overview of key attributes such as mean, standard deviation, and range, which help in understanding the spread and quality of the data. Graph 1 illustrates the detection of missing values across different features, ensuring the completeness and reliability of the dataset before further processing. Figure 5 displays the confusion matrix, which highlights the performance of the classification model by comparing actual versus predicted outcomes. Figure 6 shows the correlation matrix, enabling identification of relationships among variables and helping in feature selection. Figure 7 visualizes the distribution of the target variable, giving insights into the balance of the dataset. Graph 2 depicts the presence of heart disease based on gender, highlighting gender-wise disparities in prevalence. Graph 3 represents the distribution of the target variable in the form of a pie chart for better visualization of proportions. Graph 4 presents the same distribution through a bar chart, making it easier to compare frequencies across classes. A variety of algorithms and datasets have been explored, each demonstrating varying degrees of success based on the methods and metrics applied. In this study, we evaluate a selection of machine learning algorithms using the dataset. The performance of each technique is assessed using multiple evaluation metrics to determine their effectiveness in predicting heart disease outcomes.

8. Future Enhancement

Recent studies have explored different ways to cure heart disease. Techniques such as ensemble learning, feature extraction, and deep learning have helped researchers make prediction models more precise and effective. Looking ahead, the system could be expanded to include more advanced prediction models, tools for calculating risk, and improved feature extraction methods that can also be applied to other medical conditions. An additional improvement could be enabling the system to process and predict results for all patient records at once, making it faster and more efficient. It is also important to note that major risk factors for heart disease, such as clinical data, lifestyle, and the impact of environment, often develop without noticeable early symptoms. By closely tracking and managing these indicators, the prediction system can provide clearer and more accurate assessments of a patient's heart disease risk.

References

Ahmad, I., Alqarni, M. A., Almazroi, A. A., and Tariq, A. (2020). Experimental evaluation of clickbait detec- tion using machine learning models. *Intell. Autom. Soft Comput.*, 26: 1335–1344. doi: 10.32604/iasc.2020.013861

Ahmad, I., Rehman, S. U., Khan, I. U., Ali, A., Hussain, H., Jan, S. et al. (2021). A hybrid approach for automatic aorta segmentation in abdominal 3D CT scan images. *J. Med. Imaging Health Inf.*, 11: 712–719. doi: 10.1166/jmihi.2021.3364

Budholiya, M., Shrivastava, S. K., and Sharma, V. (2025). Advanced hybrid machine learning model for accurate detection of cardiovascular disease. *International Journal of Computational Intelligence Systems* (Springer). DOI as per source retrieval.

Cardiovascular Diseases, 2021. Available from:https://www.who.int/health-topics/cardiovascular diseases.

Chen, Y., Qin, X., Zhang, L., and Yi, B. (2020). A novel method of heart failure prediction based on DPCNN XGBoost model. *Comput. Mater. Con.*, 65: 495–510. doi: 10.32604/cmc.2020. 011278

Chicco, D., and Jurman, G. (2020). Machine learning can predict survival of patients with heart failure from serum creatinine and ejection fraction alone. *BMC Med. Inf. Decis. Making*, 20: 1–16. doi: 10.1186/s12911-020-1023-5

Ekle, F. A. et al. (2024). Machine Learning models for heart disease prediction and dietary lifestyle change therapy recommendation: a systematic review. *Discover Artificial Intelligence*, vol. 4, art. 113. DOI: 10.1007/s44163-024-00181-w

Freeman, B. S., Taylor, G., and B. Gharabaghi, J. (2018). Th´e, Forecasting air quality time series using deep learning. *J. Air Waste Manage. Assoc.*, 68: 866–886. doi: 10.1080/10962247.2018. 1459956.

Ghosh, P., Azam, S., Jonkman, M., Karim, A., Shamrat, F. J. M., Ignatious, E. et al. (2021). Effi- cient prediction of cardiovascular disease using machine learning algorithms with relief and LASSO feature selection techniques. *IEEE Access*, 9: 19304–19326. doi: 10.1109/AC- CESS.2021.3053759

Guo, C. et al. (2025). Interpretable artificial intelligence model for predicting heart failure (HF) severity using multidimensional clinical data, (*likely BMC or Springer journal*, 2025). DOI as per source retrieval

Litjens, G., Kooi, T., Bejnordi, B. E., Setio, A. A. A., Ciompi, F., Ghafoorian, M. et al. (2017). A survey on deep learning in medical image analysis. *Med. Image Anal.*, 42: 60–88. doi: 10.1016/j. media.2017.07.005.

Liu, X., Xie, Z., Zhang, Y. et al. (2024). Machine learning for predicting in-hospital mortality in elderly patients with heart failure combined with hypertension: a multicenter retrospective study. *Cardiovascular Diabetology*, vol. 23, art. 407. DOI: 10.1186/s12933-024-02503-9

Mohan, S., Thirumalai, C., and Srivastava, G. (2019). Effective heart disease prediction using hybrid machine learning techniques. *IEEE Access*, 7: 81542–81554. doi: 10.1109/ACCESS.2019.2923707

Monti, C. B., Codari, M., van Assen, M., De Cecco, C. N., and Vliegenthart, R. (2020). Machine learning and deep neural networks applications in computed tomography for coronary artery disease and myocardial perfusion. *J. Thorac. Imaging*, 35: S58–S65. doi: 10.1097/RTI.000000 0000000490

Munawar, M., and Noreen, I. (2021). Duplicate frame video forgery detection using siamese-based RNN. *Intell. Autom. Soft Comput.*, 29: 927–937. doi: 10.32604/iasc.2021.018854

Porum, M., Iadanza, E., Massaro, S., and Pecchia, L. (2020). A convolutional neural network approach to detect congestive heart failure. *Biomed. Signal Process. Control*, 55: 101597. doi: 10.1016/j.bspc.2019.101597

Rahimi, K., Bennett, D., Conrad, N., Williams, T. M., Basu, J., Dwight, J. et al. (2014). Risk prediction in patients with heart failure: a systematic review and analysis. *JACC Heart Failure*, 2: 440–446. doi: 10.1016/j.jchf.2014.04.008

Tripoliti, E., Papadopoulos, T., Karanasiou, G., Naka, K., and Fotiadis, D. (2017). Heart failure: Diagnosis, severity estimation and prediction of adverse events through machine learning techniques, Comput. *Struct. Biotechnol. J.*, 12: 26–47. doi: 10.1016/j.csbj.2016.11.001.

www.kaggle.com

Zhang, Z., Qiu, Y., Yang, X., and Zhang, M. (2020). Enhanced character-level deep convolutional neural networks for cardiovascular disease prediction. *BMC Med. Inf. Decis. Making*, 20: 1–10. doi: 10.1186/s12911-020-1118-z

Zhou, C., Dai, P., Hou, A. et al. (2024). A comprehensive review of deep learning-based models for heart disease prediction. *Artificial Intelligence Review*, vol. 57, art. 263. DOI: 10.1007/s10462-024-10899-9

Chapter 6

The Cloud–Edge Continuum for Scalable and Responsive Healthcare Delivery

Hemant Kumar Saini and *Aditya Dev Mishra**

1. Introduction

While delivering services to patients, many issues are arising. If checked with the centralised healthcare, then many challenges related to latency, scalability, and bandwidth occur. Wearable remote patient monitoring (RPMS) devices and the Internet of Medical Things (IoMT) generate a large volume of data that needs to be processed in real time (Wang et al. 2022). In the absence of an integrated framework, healthcare units also inefficiently manage local and centralised resources for anomaly detection. Due to the main issues of computational bottlenecks, latency, and bandwidth-intensive care unit monitoring, real-time anomaly detection is essential. Furthermore, there are serious bandwidth limitations due to the exponential rise of IoMT devices, which are expected to produce more than 85% of healthcare data by 2026.

Scalability is one of the major concerns faced during crises such as the COVID pandemic due to a sudden spike in patient data. Privacy concerns are another challenge, as transmitting sensitive medical data across networks increases the risk of breaches. Therefore, the integration of edge computing with cloud computing guarantees real-time responsiveness and scalability. Realising the need and significance of edge computing for healthcare patient-centric data processing and cloud services for storage and analytics, the study of the proposed chapter is as follows: "*The Cloud–Edge Continuum for Scalable and Responsive Healthcare Delivery*". This study has been beneficial for researchers, students, and academics who work in the field of cloud-edge-related services in the healthcare sector.

School of Computer Science Engineering & Technology, Bennett University, Greater Noida, India.
Email: hemantrhce@rediffmail.com, https://orchid.org/0000-0002-2759-2654
* Corresponding author: aditya2aug@gmail.com, https://orchid.org/0000-0002-7060-2600

DOI: 10.1201/9781003783855-8

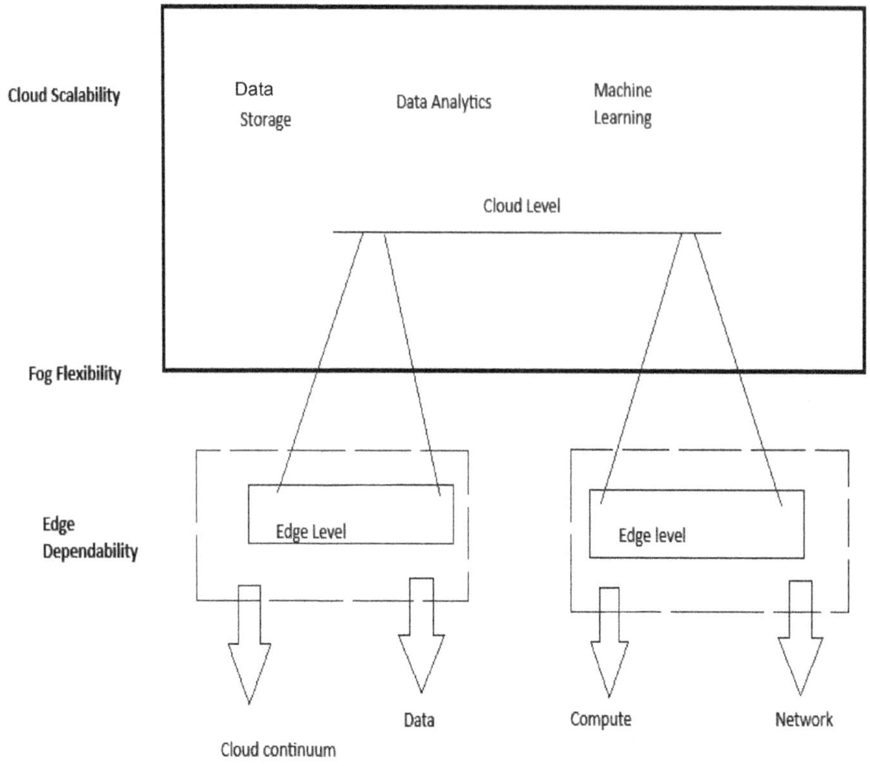

Figure 1: Cloud–edge continuum architecture

3.2 Real-Time Requirements in Critical Care

Latency is not only annoying in many healthcare settings, but it can be fatal, particularly in critical care and emergency medicine. Sub-second reaction speeds are crucial for applications such as AI-powered stroke diagnosis, remote robotic surgery, and automated defibrillator activation. Delays of just a few hundred milliseconds can jeopardise patient safety or lower intervention success rates. Round-trip latency in traditional cloud-based architectures can range from 50 to 200 milliseconds, depending on the location and network circumstances, as data must frequently transit across lengthy network links to centralised data centres. This is insufficient for time-sensitive physiological monitoring, but it might be fine for administrative duties like scheduling or billing (Nezami et al. 2021).

By locating computational resources closer to the data centre—whether in an ambulance's onboard processor, a hospital's local server cluster, or even on-device AI chips—edge computing helps to overcome this issue. Healthcare systems can almost immediately initiate alerts or therapeutic treatments by assessing respiratory distress patterns or ECG anomalies at the edge, while the cloud is still in charge of long-term analytics, storage, and model retraining. Telestroke programmes, in which neurologists evaluate patients remotely via real-time video feeds and imaging, provide a practical example. By incorporating edge-based AI models, quick preliminary diagnosis is made possible, guaranteeing that thrombolytic therapy door-to-needle timeframes are kept to a minimum.

3.3 Security, Privacy, and Compliance (HIPAA, GDPR)

One of the most sensitive types of personal data is healthcare data, which is subject to stringent laws like the GDPR (General Data Protection Regulation) in the EU and HIPAA (Health Insurance Portability and Accountability Act) in the US. Due to mobile patient data running over public network data centres, sharing requires numerous ethics and consents in collection and availability. Unauthorised access, data breaches, and non-compliance with jurisdiction-specific data localisation regulations are all made more likely by this.

By allowing local data retention for sensitive aspects and encrypted aggregation for cloud-based analytics, the cloud–edge continuum provides a compliant pathway. For instance, before being uploaded to cloud storage, raw medical images might be anonymised at the edge by eliminating facial features or personal identifiers under the GDPR's "data minimisation" principle. Similarly, HIPAA-covered organisations can still leverage multi-institutional datasets and maintain privacy by using federated learning to train AI diagnostic models without sending raw patient data (Shi et al. 2016). Nevertheless, compliance is not always guaranteed by cloud and edge integration. It is necessary to implement zero-trust network topologies, role-based access controls, and ongoing security audits. In order to manage cross-border data flows while providing services to patients abroad or running telemedicine platforms, healthcare providers also require technical and legal harmonisation.

3.4 Opportunities Enabled by Cloud–Edge Integration

The cloud–edge continuum opens up revolutionary possibilities for scalable and responsive healthcare delivery, notwithstanding the substantial obstacles. Manually, doctors might take actions against critical cases and give immediate treatments, but decision-making in real time can be more effective if implemented for seizure alerts, using long-term patterns on cloud (Shi et al. 2016).

3.5 Scalable Telemedicine Services

Through tele-consultations, a new strategy was applied during pandemic times for online consulting through doctors, but it created a new challenge when large number of patients came. It became scalable by uploading the data to the cloud at low-latency video.

3.6 Improved Medical Imaging Workflows

With edge-based AI algorithms, images are segmented, which not only reduces the work of radiologists but also enhances their accuracy by training models on various datasets.

3.7 Interoperable Health IoT Ecosystems

Electronic health records (EHR) transform various medical devices' data into standardised formats (such as HL7, FHIR). Additionally, this avoids lock-in and enhances vendor neutrality.

3.8 Resource Optimisation in Hybrid Environments

With the batch processing era, medical data is processed dynamically by switching between cloud and edges, such as AI inference at the edge (Huamei Qi et al. 2024). With this hybrid cloud–edge continuum, healthcare has transformed into digital, advanced patient-centric systems.

4. Fundamentals of the present study

The core elements of the cloud–edge continuum are thoroughly examined in this section, including edge computing, cloud consortiums in healthcare, fog computing, and crucial enabling technologies like IoT, AI, and 5G.

4.1 Edge Computing in Healthcare

One such advancement is seen in edge computing in smart traffic driverless cars (Lilhore et al. 2022), where decentralised cars save energy, but some data at the edges, for their fastness, are prone to cyber challenges.

4.2 Cloud Consortium in Healthcare

A new topological approach integrates cloud and fog over 5G and software-defined networking (SDN) in a systematic architecture from local computers to fog computers that improves bandwidth efficiencies.

4.3 Key Enablers: AI, IoT, and 5G

Artificial intelligence (AI), Internet of Things (IoT), and 5G networks are intelligent diagnostic capabilities of the cloud–edge continuum in healthcare through which to handle environmental, behavioural, and physiological data in real time smartly. With these key enablers, medical data is updated with AI models by federated learning that are deployed at the edges in ultra-low latency urgently without interruption. Therefore, through these key enablers, next-generation healthcare delivery systems achieve scalability and reactivity seamlessly.

5. Cloud–Edge Architecture for Healthcare Delivery

In order to maximise resource management, interoperability, and performance, a cloud–edge architecture for healthcare delivery uses a tiered approach that integrates devices, edge computing, fog computing, and cloud computing. This architecture supports real-time analysis, effective data processing, and secure data storage for a wide range of healthcare applications.

5.1 An Overview of Prescriptive Architecture

The prescriptive architecture of the cloud–edge healthcare system comprises four distinct layers: the cloud layer, fog layer, edge layer, and device layer, as illustrated in Figure 2. The prescriptive architecture is a bottom-up approach that starts from the device layer and ends with the cloud layer.

As depicted in Figure 2, a prescriptive system architecture is the integration of layers from the medical device, including an IoT-enabled healthcare device, to the edge, fog, and cloud computing techniques.

(a) Device Layer: This is the first layer responsible for the collection of raw medical health data through tests by medical devices such as an ECG monitor, a glucometer, a blood pressure monitor, a thermometer, or an X-ray machine.

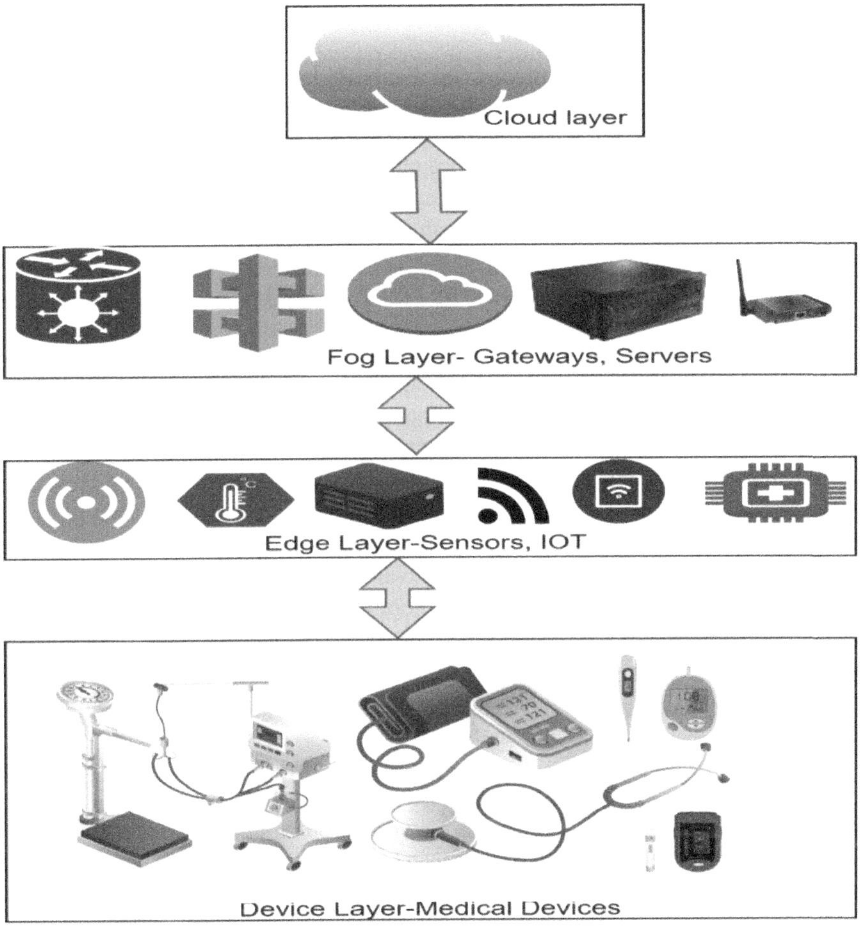

Figure 2: System architecture of cloud-edge healthcare system.

256 or similar algorithms, along with secure communication protocols such as TLS 1.3. Lightweight encryption techniques provide a balance between performance and security for edge devices with constrained processing power. The multi-factor authentication approach in identity verification and access control systems for sensitive operations and attribute-based access control (ABAC) systems are equally critical. Automated certificate rotation, hardware-based trusted platform modules (TPM), and secure device onboarding procedures further strengthen defence against unauthorised intrusion. The final aim is to achieve a layered defence system that preserves confidentiality, integrity, and availability over the whole cloud–edge data flow.

7.2 Privacy-Preserving Computation (Federated Learning, Homomorphic Encryption)

In regular industries that have critical infrastructure, such as healthcare and finance, the problem of extracting value from dispersed data without disclosing personal information is addressed by privacy-preserving computation. Federated learning is based on privacy-preserving computation because it trains an AI model directly on edge-based decentralised datasets, sharing only model parameters with the cloud rather than raw data. Therefore, it reduces the dangers related to central data aggregation and also facilitates collaborative intelligence. Similarly, homomorphic encryption (HE) executes calculations directly on encrypted data and ensures that no node can see the underlying data, becoming more popular due to its increased efficiency. An additional layer of protection provides privacy by adding statistical noise to datasets or AI outputs that keeps people from being re-identified even when datasets are repeatedly queried. Secure multi-party computation (SMPC) encourages cross-entity collaboration without sacrificing privacy by allowing many organisations to analyse encrypted datasets together without disclosing their inputs.

7.3 Regulatory Compliance and Auditing

In the cloud–edge continuum, compliance with legal frameworks that control data processing, handling, and storage must be combined with security and privacy protections. The standard guidelines on data protection requirements are established by regulations such as the GDPR in Europe, HIPAA in the United States, and ISO/IEC 27001 standards for information security. These regulations affect architectural design, cross-border data transfers, and sovereignty requirements that specify the actual location of data. Organisations are increasingly using compliance-as-code approaches, integrating regulatory checks into automated workflows and orchestration tools to ensure ongoing alignment with these legal requirements.

Auditing procedures, including real-time log gathering, immutable blockchain-based audit trails, and third-party certification procedures such as the Cloud Security Alliance STAR programme, are mandatory to preserve confidence and transparency. The improvement of compliance follows the principle of verification at every stage of access and communication, and the implementation of Zero Trust Architecture (ZTA) holds that no person, device, or application is intrinsically trusted.

8. Use Cases and Case Studies

This section provides use cases and case study examples illustrating the role of the cloud–edge continuum in enabling scalable and responsive healthcare delivery systems.

8.1 Remote Patient Monitoring and Telemedicine

Improving Chronic Care Delivery Through Telehealth and Remote Patient Monitoring: A Case Study of Community Nurse Home Care (2022–2023): This case study described the implementation and management of a telehealth and remote patient monitoring (RPM) programme by a non-profit home health and hospice provider, Community Nurse Home Care (CNHC) in Massachusetts. One of the initiatives of this group, which ran from December 2022 to December 2023, used Health Recovery Solutions' PatientConnect® Complete to monitor patients with chronic conditions such as CHF, COPD, diabetes, and hypertension through the integration of daily biometric tracking, symptom surveys, educational videos, and centralised telehealth coordination (Shuaib et al. 2023, Vudathaneni et al. 2024). The objective of this programme was to improve patient involvement, encourage self-management, and stop preventable acute care events by focusing on high-acuity patients who had previously been hospitalised or who did not comply (healthrecoverysolutions.com). With 96% of patients reporting greater involvement in their care and 93% saying they were better supported by their healthcare team, the programme achieved an average adherence rate of 88% throughout 2023. Notably, telehealth participants saw a 17% decrease in average 60-day acute care hospitalisation (ACH) rates from January 2022 to December 2023 as compared to non-enrolled patients. The decrease was 23% among patients with intermediate risk (SHP risk scores 2–4).

8.2 AI-Assisted Diagnostics at the Edge

From 2022 to 2030, the global surgical robotics market is expected to expand at a compound annual growth rate (CAGR) of 19.3% due to developments in artificial intelligence (AI), surgical navigation, and medical imaging that increase accuracy, lower mistakes, and improve clinical outcomes. The use of the Advantech EPC-B5587 embedded PC as the main computer platform for a neurosurgery robot driven by artificial intelligence is examined in this case study. The integration of CT/MRI fused imaging and brain mapping provides real-time intraoperative tasks more easily and improves the AI paradigm's imaging executions at the edge level. For instance, Dual Gigabit Ethernet ensures zero-latency data transfer, while the EPC-B5587's 10th Gen Intel® Xeon® CPU with NVIDIA® Quadro® GPU provides high-performance computation. Such a system helps in enhancing diagnosis and benefits healthcare.

9. Conclusion

The new advancement of the consortium merges all the services of fog, edge, and cloud to transform the healthcare service delivery system. Old cloud systems, though powerful in capabilities, face latencies in processing; edge systems are not scalable. The cloud–edge continuum jointly adapts both scalability and responsiveness. This chapter explores healthcare challenges and opportunities, emphasising the management of large volumes of medical device data, real-time critical care requirements, and compliance with security, privacy, HIPAA, and GDPR. This chapter also demonstrates opportunities such as scalable telemedicine services, interoperable health IoT ecosystems, and resource optimisation in hybrid environments enabled by cloud–edge integration. This chapter also discusses the overview of cloud–edge architecture for healthcare delivery. This chapter examines the scalability, responsiveness, security, privacy, and trust within the cloud–edge continuum. This chapter also examines a use case and a case study that demonstrate the advantages of cloud–edge integration in healthcare systems.

References

Balouek-Thomert, D., Renart, E. G., Zamani, A. R., Simonet, A., and Parashar, M. (2019). Towards a computing continuum: Enabling edge-to-cloud integration for data-driven workflows. *The International Journal of High Performance Computing Applications,* 33(6): 1159–1174. https://doi.org/10.1177/1094342019852126

Belcastro, L., Marozzo, F., Orsino, A., Talia, D., and Trunfio, P. (2023). Edge-cloud continuum solutions for urban mobility prediction and planning. *IEEE Access,* 11: 38864–38874. https://doi.org/10.1109/ACCESS.2023.3266472

Bernabé, I., Fernández, A., Billhardt, H., and Ossowski, S. (2022, July). Towards semantic modelling of the edge-cloud continuum. In *International Conference on Practical Applications of Agents and Multi-Agent Systems* (pp. 71–82). Springer. https://doi.org/10.1007/978-3-031-16112-3_6

Habibi, P., Farhoudi, M., Kazemian, S., Khorsandi, S., and Leon-Garcia, A. (2020). Fog computing: A comprehensive architectural survey. *IEEE Access,* 8: 69105–69133. https://doi.org/10.1109/ACCESS.2020.2983253

Huamei Qi, Fang Ren, Leilei Wang, Ping Jiang, Shaohua Wan, and Xiaoheng Deng. (2024). Multi-Compression Scale DNN Inference Acceleration based on Cloud-Edge-End Collaboration. ACM Trans. Embed. Comput. Syst. 23, 1, Article 16 (January 2024), 25 pages. https://doi.org/10.1145/3634704

Li, E., Zeng, L., Zhou, Z., and Chen, X. (2019). Edge AI: On-demand accelerating deep neural network inference via edge computing. *IEEE Transactions on Wireless Communications,* 19(1): 447–457. https://doi.org/10.1109/TWC.2019.2948610

Lilhore, U. K., Simaiya, S., Garg, A., Verma, J., and Garg, N. B. (2022). An efficient energy-aware load balancing method for cloud computing. In *2022 4th International Conference on Artificial Intelligence and Speech Technology (AIST)* (pp. 1–5). IEEE. https://doi.org/10.1109/AIST54884.2022.00007

Mehran, N., Kimovski, D., and Prodan, R. (2021). A two-sided matching model for data stream processing in the cloud–fog continuum. In *2021 IEEE/ACM 21st International Symposium on Cluster, Cloud and Internet Computing (CCGrid)* (pp. 514–524). IEEE. https://doi.org/10.1109/CCGrid51090.2021.00061

Nezami, Z., Zamanifar, K., Djemame, K., and Pournaras, E. (2021). Decentralized edge-to-cloud load balancing: Service placement for the Internet of Things. *IEEE Access,* 9: 64983–65000. https://doi.org/10.1109/ACCESS.2021.3075749

Shi, W., Cao, J., Zhang, Q., Li, Y., and Xu, L. (2016). Edge computing: Vision and challenges. *IEEE Internet of Things Journal,* 3(5): 637–646. https://doi.org/10.1109/JIOT.2016.2579198

Shuaib, M., Bhatia, S., Alam, S., Masih, R. K., Alqahtani, N., Basheer, S. et al. (2023). An optimized, dynamic, and efficient load-balancing framework for resource management in the Internet of Things (IoT) environment. *Electronics,* 12(5): 1104. https://doi.org/10.3390/electronics12051104

Vudathaneni, V. K. P., Lanke, R. B., Mudaliyar, M. C., Movva, K. V., Kalluri, M., and Boyapati, R. (2024). The impact of telemedicine and remote patient monitoring on healthcare delivery: A comprehensive evaluation. *Cureus,* 16(3): e55534. https://doi.org/10.7759/cureus.55534

Wang, Y., Su, Z., Zhang, N., Xing, R., Liu, D., Luan, T. H. et al. (2022). A survey on metaverse: Fundamentals, security, and privacy. *IEEE Communications Surveys & Tutorials,* 25(1): 319–352. https://doi.org/10.1109/COMST.2022.3144724

Yi, S., Hao, Z., Qin, Z., and Li, Q. (2015). Fog computing: Platform and applications. In *2015 Third IEEE Workshop on Hot Topics in Web Systems and Technologies (HotWeb)* (pp. 73–78). IEEE. https://doi.org/10.1109/HotWeb.2015.22

been evident in radiation therapy planning, surgical planning optimization, and resource optimization of a complicated healthcare delivery system.

The confluence of quantum computing with IoMT infrastructure in smart hospitals offers unparalleled insights into real-time analysis of complex health data that is unable to be analyzed currently because of computational constraints (Rajawat et al. 2023). Edge computing devices based on quantum mechanics would allow analysis of real-time physiological data streams provided by a variety of IoMT sensors, identifying critical trends and anomalies that may indicate an imminent emergency situation prior to developing clinical symptoms. Still, the implementation of quantum computing into an array of healthcare systems does cause a number of issues related to compatibility between the systems, data standardization, and the necessity of a quantum–classical hybrid architecture that will allow the use of existing IT systems in healthcare while at the same time providing the systems with quantum computing capabilities (Tsuchiya et al. 2024).

1.3 Importance of Privacy and Ethics in Next-Gen Healthcare

The change in healthcare science using quantum computing and IoMT technologies requires a rethinking of the privacy protection process and ethical codes of conduct concerning the use of medical data (Jeyaraman et al. 2024). Healthcare data is one of the most sensitive types of personal data, being not only a report on the present situation but also a red flag of the future health state, hereditary factors, and sexual peculiarities of behavior that could be used against the owner. By incorporating the ability to conduct quantum computation into the system in which medical data is processed, both the possible benefits of the application and the risks it poses increase, given the potential of such an approach to eliminate the current method of protection (the risks) and to provide hitherto inconceivable means of encryption (the benefits) of data.

The ethical concerns regarding the quantum-enhanced healthcare are not confined to the technological issues of data security but entail deeper inquiry regarding patient rights, autonomy and informed consent, and the distribution of advanced medical techniques responsibly. As the sophistication of predictive analytics and automated decision-making systems increases at the scale made potentially possible by quantum computing, there is growing concern over the transparency of algorithms, the possibility of biases being built into systems, and accountability measures taken when quantum-enhanced AI-based systems make critical medical decisions. The fact that quantum computers are capable of cracking the most current methods of encryption presents obvious risks to current healthcare datasets, which would allow unauthorized access to decades of gathered medical information.

Data privacy issues in quantum medical care settings need to combat a variety of privacy-related challenges, such as data security in the IoMT transfer environment, data security on medical records when targeted by future quantum hacking, and patient anonymity in medical research sets subject to future quantum-powered identification (El Emam et al. 2011). The use of data minimization becomes more

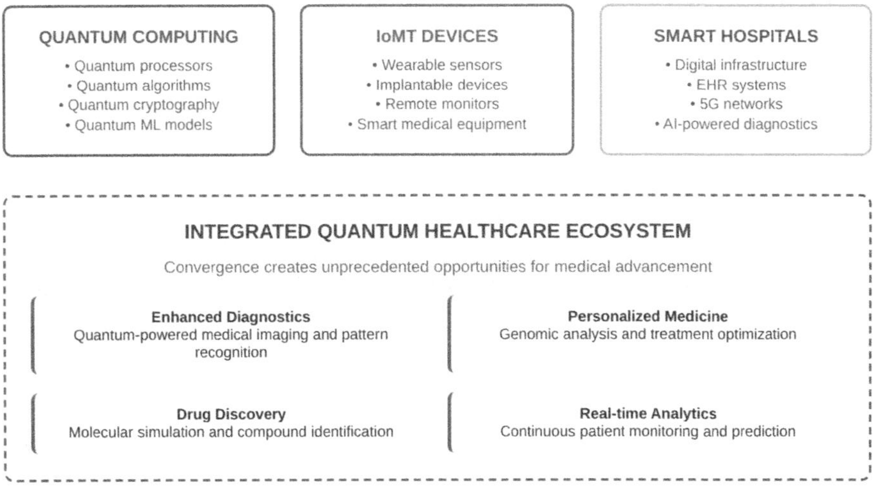

Figure 1: Quantum Healthcare Ecosystem Overview.

complicated when quantum analytics can cull out information previously not possible, given innocent-looking pooling or anonymizing of data, a pooling that forms sensitive health knowledge. The alignment of quantum computing with IoMT devices and smart hospital infrastructure creates a powerful yet challenging healthcare environment that is shown in Figure 1.

The quantum healthcare ethics will have to strike a balance between the enormous potential of medical inventions on the one hand and the dangers of developing surveillance-system-friendly health monitoring systems on the other hand that will pose possible threats to the freedom and privacy of individuals. The main limitation is that, even though quantum-resistant cryptographic protocols are necessary to secure healthcare data, they have to be developed in a manner that does not disrupt the interoperability of the system and the flow of emergency medical treatment. In addition, cross-border data flow and the internationalization of quantum computing development make it imperative to undertake international cooperation in the setup of both an ethical framework and data privacy regulations that uphold the rights of patients whilst allowing health-related research to take place.

The concept of informed consent, as now comprehended, requires revision in the context of quantum healthcare due to the complexities of quantum algorithms and the probabilistic character of quantum calculations, which can entail results that patients can neither comprehend nor anticipate. Healthcare institutions will need to come up with new approaches to patient engagement that ensure openness concerning the use of quantum computing applications, yet avoid technical jargon that would deter patient willingness to sign up for participatory quantum-enhanced medical programs that promise emancipating benefits. It is imperative to develop ethical review boards educated on quantum computing implications to review research protocols and clinical applications that build upon quantum capabilities.

2. Data Privacy in Quantum Healthcare

2.1 Understanding Healthcare Data Sensitivity (EHRs, IoMT, Genomic Data)

The nature of healthcare data is unique in that it is sensitive, permanent, and can provide intimate information about a person and his or her physical and mental state (Yi et al. 2022). Electronic Health Records (EHRs) represent the totality of medical records of each patient, which include diagnosis findings, treatment plans, pharmaceutical information, surgical operations, and psychosocial tests that combine to form disease-related timelines of individuals. The privacy of EHR data is not only significant to medical purposes but also to job opportunities, insurance provisions, social life, as well as personal image, since when shared in the wrong context it may impact negatively on all these aspects (Banerjee et al. 2023).

IoMT integration has led to an explosion in the volume and resolution of health data collection such that there are now persistent streams of measurements recording patients' physiological performance, their behavioral patterns, and interactions with the environment, providing information on patient daily living activities of unprecedented richness. Wearable sensors include heart rate variability, sleep reports, physical activity, and stress biometrics, and have the ability to compile comprehensive behavioral profiles that are beyond the potential of clinical observations. The difficulties encountered in the treatment and monitoring of an individual are expected in implantable medical devices like cardiac pacemakers, continuous glucose monitors, and neural stimulators, as highly sensitive medical data can be generated regarding organ functionality as well as neurological activity. This can reveal not only the current state of health but also predict future medical states with higher accuracy.

Genomic data, potentially the most sensitive form of health information, consists of fixed genetic strings that determine disease susceptibility or drug metabolism patterns and that affect not only the individual but also his or her biological family. Analysis of genomic data using quantum computing programs allows determination of complex polygenic predisposing factors, rare genetic changes, and epigenetic shifts contributing to the appearance of diseases and the effectiveness of treatment. In contrast to other medical information, which reflects time variations of health, genomic information cannot be altered even when it has been compromised, leaving the individual exposed to risks of permanent privacy damage in this data type.

2.2 Role of IoMT Devices in Data Collection and Risks

The integration of IoMTs into healthcare has resulted in total monitoring environments within which physiological, environmental, and behavioral factors are measured with a continuity and accuracy heretofore unseen (Hanji 2020). Such devices include consumer-level health-monitoring watches to high-quality medical niche equipment, and they comprise an estimated 2.5 quintillion bytes per day of data across universal healthcare systems. This extensive IoMT growth in smart hospital settings generates a complete patient monitoring system where several sensors monitor vital signs, medication compliance, movement, and recovery rates of the patient at a time.

IoMT networks have a distributed architecture, which presents numerous weaknesses at which vulnerable health data could be intercepted, edited, or stolen by an attacker. Every IoMT device poses a possible threat of cyberattacks, and many have limited security features because of processing power, battery life constraints, and cost. The non-homogeneous structure of IoMT environments, with devices produced by different companies and operating at different levels of security, opens potential risks to unbalanced protection in the healthcare data collection system.

Certain security issues arise with wireless communication protocols used in IoMT to transmit data, since wireless signals can be intercepted, jammed, or even spoofed without having to physically access the device. The connectivity of the Internet of Medical Things (IoMT) through Bluetooth, Wi-Fi, and cellular networks poses risks such as eavesdropping, man-in-the-middle attacks, and denial-of-service attempts. These vulnerabilities can compromise data confidentiality and device recoverability, exposing the information exchanged among medical devices. The potential for attacks known as man-in-the-middle further jeopardizes the availability of data between connected devices, leading to the disclosure of transmitted patient information (Hasan et al. 2021).

2.3 Quantum Computing's Potential in Healthcare Data Processing

Quantum computing introduces transformative capabilities in the processing of healthcare data far beyond classical computational means, more specifically in the analysis of high-dimensional medical data (Huang et al. 2021). The quantum mechanical principle of superposition allows quantum processors to computationally explore a number of data configurations all at once, producing exponentially faster performance on pattern recognition applications essential to diagnosis and treatment optimization efforts. Quantum algorithms have been shown to be especially effective with processing unstructured medical information such as natural language clinical notes, medical imaging, and continuous physiological signals provided by IoMT devices.

Quantum machine learning algorithms' potential applied to patient care in such health institutions includes the possibility to identify invisible correlations and nonlinearities that classical analysis tools do not reveal. Quantum support vector machines and quantum neural networks have demonstrated the capability to outperform classical methods in classifying complicated disease conditions, predicting treatment responses, and identifying novel markers on the basis of multi-topic datasets (Sharma et al. 2023). The high-feature-space computational ability of quantum computers makes them ideal for genomic analysis, where thousands of genetic variants interact in influencing susceptibility to disease or drug metabolism.

Quantum optimization algorithms present an excellent example of groundbreaking approaches to healthcare resource safety, treatment planning, and clinical trial design, as they are able to evaluate enormous solution spaces that would take classical computers millennia to work through. The use of quantum annealing in radiation therapy planning has shown enhanced beam settings to optimize tumor algorithms and the idiosyncrasies of quantum mechanics could complicate situations where implementation errors have a chance to compromise healthcare data security. The limited availability of quantum computational resources requires the external procurement of quantum computers through quantum processing cloud services, which by extension poses a further risk to the privacy of sensitive healthcare data through external sharing with a third party.

3. Ethical Considerations in Quantum-Enabled Healthcare

3.1 Patient Rights, Consent, and Data Ownership

The impact of quantum computing on the healthcare system dislodges the existing notion of patient rights when incorporated into the systems fundamentally, as it warrants a complete revisiting of autonomy, consent, and data ownership constructs (Pruski 2024). The potential of quantum computing to draw out unparalleled insights in healthcare data preconditions the emergence of cases where the health information disclosed within a single medical claim could reveal unrelated healthcare issues, genetic predispositions, and behavioral patterns without the patient's knowledge or consent. The informed consent principle becomes even more complicated as quantum algorithms allow inferring probabilistic health inferences from seemingly unrelated datasets, and it will hardly be comprehensible to the patient what his or her medical data might be used to do.

Quantum data ownership goes beyond mere possession by also referring to ownership of quantum states, as well as ownership of entangled information and its inferred insights, which may have independent value. When patient data are put through quantum processing, they are also held in superposition states, which challenge classical ideas of discrete ownership of data, potentially creating issues over who owns quantum-derived insights and whether patients have any rights to them (David 2025). Quantum fingerprints or quantum signatures made with patient data produce new forms of biometric information that might last forever and can be misused in unexpected ways.

Patient autonomy in the quantum-enhanced healthcare system will need to be combined with meaningful control over how quantum computing technologies handle personal health-related information. The right to data portability also becomes thorny when quantum processing generates entangled states that cannot be cleanly partitioned or transferred between healthcare providers, and the quantum advantages are lost. Patients' demands for the right to rectification and erasure of patient data need to take into consideration that, due to the characteristics of quantum information, it is challenging to completely erase records without leaving some remnant in the quantum dataset through error correction and decoherence.

3.2 Fairness, Transparency, and Accountability in Quantum AI Systems

Considering the ability of quantum computing in reducing pre-existing biases by virtue of its ability to process biased training data, quantum AI in healthcare settings brings up the question of algorithmic fairness (Shah et al. 2025). That is, quantum machine learning algorithms can reproduce or even escalate disparities in healthcare by training on datasets that are unrepresentative of minority populations, rare diseases, or disadvantaged groups. The superposition principle in quantum computing could create conditions where different demographic groups attain varying degrees of diagnostic accuracy or differing qualities of treatment recommendations based on quantum state preparation protocols.

Transparency in quantum healthcare AI systems is particularly problematic because quantum behavior entails the measurement problem in quantum mechanics, in which measuring quantum states collapses superposition and, depending on the computation, possibly changes the results. The opaqueness of quantum algorithms relative to classical neural networks becomes further obscured by the fact that quantum effects such as entanglement and quantum interference patterns result in decision pathways that are unobservable, since accessing them destroys the computation itself (Montanaro 2015). Quantum-based medical recommendations presented to healthcare providers and patients must have explainable AI mechanisms that enable a quantum-mechanical interface between the patient and healthcare providers.

Quantum healthcare AI accountability models should take into account the probabilistic nature of quantum programs, in which repeated runs with the same input can give different outputs depending on quantum measurement uncertainty. Because quantum algorithms may be involved in making major medical decisions, allocating liability in the event of error is complicated when the randomness of quantum algorithms plays a role in the outcome. The complex nature of the partnership between quantum hardware producers, algorithm designers, hospitals, and medical practitioners necessitates clearly defined parameters of liability and appropriate governance mechanisms.

3.3 Ethical Dilemmas in Automated Decision-Making

The implementation of quantum-enhanced automated decision-making tools in clinical practice poses severe ethical questions regarding whether computational efficiency should outweigh human medical judgment (Gomathy et al. 2024). Quantum algorithms with the potential to process multidimensional medical data that has been accumulated by the healthcare industry can discover disease trends and correlations of treatment options outside of human cognition, but their algorithms exist in the realms of probabilities, out of which some harmful effects could take place which contravene the medical ethics principles of providing benefits while avoiding harm to the patient, as embodied in the tenets of beneficence and non-maleficence. Quantum algorithms that propose the most statistically optimal interventions also present ethical dilemmas for healthcare providers when interventions that are statistically supported conflict with patient preferences, cultural beliefs, or quality of life considerations.

Even more complicated is the question of moral responsibility when quantum AI systems offer solutions not included in accepted medical practices but derived from quantum-discovered population health trends. The uncertainty of quantum prediction introduces ambiguity into life-or-death medical decision-making, making it imperative to establish new mechanisms for communicating risk and obtaining informed consent where medical prescriptions reflect probability distributions rather than determinate outcomes.

The ethics of quantum optimization algorithms in resource allocation is challenged by the tension between computational efficiency and the principles of equality and equitable rights (Bhimavarapu 2025). Quantum algorithms could optimize healthcare resource distribution in a manner that benefits the overall population while disadvantaging individuals with low probabilities of survival or those with rare diseases. The application of quantum computing to calculate cost-effectiveness in treatment procedures has the potential to discriminate against elderly, disabled, or chronically ill patients by algorithmically determining that prioritizing them is less efficient.

4. Privacy Protection Models and Ethical Frameworks

4.1 Classical vs. Quantum Encryption in Healthcare

The fundamental distinction between classical and quantum forms of encryption in medical care situations is indicative of a striking difference in the underlying mathematical and physical principals, measures of security assurance, and enforcement constraints (Saarinen 2022). Classical encryption systems rely on the computational difficulty of issues such as factoring large prime numbers or calculating discrete logarithms, which have formed the foundation of healthcare data security over the past decades. Systems, such as RSA, AES and elliptic curve, depend on the intractability of some mathematical problems on classical computers, needing exponential time in order to defeat an attempt to find a solution through brute-force search.

Quantum encryption, however, uses the laws of quantum mechanics—e.g., the no-cloning theorem, quantum entanglement, and uncertainty principle—to offer information-theoretic security that can never be broken by means of powers of computation (Bennett et al. 2023). Classical to quantum encryption change in healthcare is not just another technological enhancement; rather, it entails a paradigm shift in encryptions of medical data confidentiality. Where classical encryption can be decrypted in the future even after already encrypted data was stored and subsequently attacked on a stronger computer, quantum encryption offers forward security against all future technological advances, including fault tolerant quantum computers. Figure 3 shows an exhaustive mapping of privacy issues in quantum healthcare and their respective technical and governance solutions to show the multilayered structure that is required in order to secure their privacy.

Table 1 provides a detailed comparison between classical and quantum privacy models, highlighting the fundamental differences in security foundations, implementation requirements, and implications for healthcare data protection.

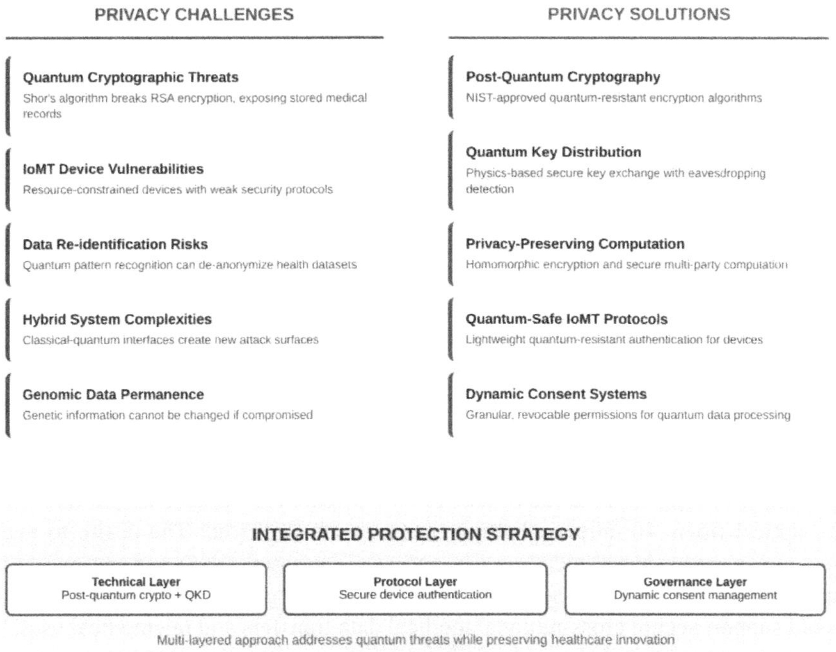

Figure 3: Privacy Challenges and Solutions in Quantum Healthcare.

Table 1: Comparison of Classical and Quantum Privacy Models.

Aspect	Classical Privacy Model	Quantum Privacy Model	Implications for Healthcare
Security Basis	Computational complexity assumptions	Fundamental physical laws	Quantum provides unconditional security
Encryption Method	Mathematical algorithms (RSA, AES)	Quantum key distribution, quantum states	Healthcare data protected against future attacks
Key Distribution	Public key infrastructure, certificates	Quantum entanglement, BB84 protocol	Automatic eavesdropping detection
Threat Resistance	Vulnerable to quantum attacks	Quantum-resistant by design	Long-term medical record protection
Implementation Cost	Low (software-based)	High (specialized hardware required)	Significant infrastructure investment needed
Scalability	Highly scalable	Limited by physical constraints	Challenges for large healthcare networks
Emergency Access	Established break-glass procedures	Quantum states cannot be copied	New emergency protocols required

vulnerabilities of all systems (Sevilla and Riedel 2020). Quantum computers that can run Shor's algorithm can break RSA-2048 encryption in about 8 hours when fault-tolerant quantum systems with about 20 million physical qubits can be realized, risking exposure of virtually all of the encrypted data in healthcare organizations that have been amassed over many decades. The time when this quantum apocalypse will take place is also still uncertain, with estimations spanning between 10 and 30 years, but the longevity of medical record data is such that quantum-resistant cryptographic protections have to be implemented immediately.

The problems of scalability in quantum healthcare systems occur in several dimensions, such as the physical scale of the quantum computer, the scale of the channel in quantum-enhanced communication networks, and the scale of hybrid classical-quantum algorithms. Existing quantum processors have poor coherence times, on the order of microseconds, limiting the complexity of the healthcare computations that may be executed before decoherence ruins quantum information (Jnane et al. 2022).

5.2 Security Threats and Ethical Risks

Quantum computing convergence with IoMT ecosystems brings an entirely new level of security threats to the realm of healthcare that cannot be easily handled by existing paradigms of healthcare cybersecurity. Skilled attackers can still crack even quantum security systems, as has been seen in quantum hacking techniques such as photon-number-splitting attacks on quantum key distribution systems and side-channel attacks against quantum computers (Papastergiou et al. 2020). Quantum healthcare spaces extend the potential impact of IoMT device vulnerabilities in that quantum attacks on compromised devices have the potential to inject malicious quantum states or manipulate quantum measurements to corrupt medical data or decision-making activities.

Amplificatory effects of quantum computing on algorithmic bias pose an excessive ethical risk, as it has the potential to enhance and deepen healthcare disparities on unprecedented levels. Quantum machine learning algorithms trained over historically biased medical data may learn inherited discrimination patterns in quantum superposition states, causing such bias to be exponentially harder to detect and fix than when using classical systems (Swaminathan and Danks 2024). The broad overview of the challenges affecting quantum healthcare implementation is presented in Table 2, where the focus is put on describing the challenges by type, their impact on healthcare overall, the complexity of their alleviations, and the expected deadlines of their elimination or improvement.

6. Future Directions for Secure and Ethical Quantum Healthcare

6.1 Policy Implications: Governance and Global Standards

Generating structured governance infrastructures in quantum healthcare is unprecedented and needs exceptional coordination of technological, medical, ethical, and regulatory stakeholders to tackle the novel issues presented by quantum

Table 2: Summary of Key Challenges in Quantum Healthcare Privacy.

Challenge Category	Specific Issues	Impact Level	Mitigation Complexity	Timeline for Resolution
Quantum Attacks	Shor's algorithm breaking RSA	Critical	High	5–10 years
Scalability	Limited qubit coherence times	High	Very High	10–15 years
Integration	Classical-quantum interface	High	Medium	3–5 years
IoMT Vulnerabilities	Resource-constrained security	High	Medium	2–5 years
Algorithmic Bias	Quantum amplification of discrimination	Critical	Very High	5–10 years
Access Inequality	Cost and infrastructure barriers	High	High	10–20 years
Regulatory Gaps	Absence of quantum-specific laws	Critical	Medium	2–5 years

computing in clinical contexts (Dhingra and Dabas 2020). The formulation of policy should look ahead at how rapidly the field of quantum is developing and breeding systems and be able to adopt those advances, yet still maintain core sets of privacy and ethical standards. Such national quantum strategies are increasingly taking healthcare as an important application field, with governments committing billions to quantum medical research whilst also developing regulatory guidelines to guide the deployment of healthcare applications of quantum.

6.2 Integration of Blockchain and Quantum Technologies

The recent development of blockchain technology together with quantum technologies promises synergetic potential, providing complementary strengths in distributed trust and quantum-resistant cryptography to increase healthcare data security (Nimbe et al. 2022).

New quantum blockchains with quantum security can serve as audit records of quantum healthcare activities and defend against both classical and quantum attacks on data integrity. Post-quantum blockchain-based protocols, which specifically consider the needs of a healthcare scenario, are needed, particularly with regard to privacy requirements, selective disclosures, and emergency access.

6.3 Toward Sustainable and Inclusive Quantum Healthcare Models

Quantum computing infrastructures demand a great deal of energy and have a significant environmental footprint, which is why sustainable quantum healthcare ecosystems will need to consider the energy efficiency of their quantum computing infrastructure. Existing dilution refrigerators serving superconducting quantum computers require about 25 kilowatts of power at all times and thus present large carbon footprints to quantum healthcare buildings. Quantum healthcare that is inclusive should be aggressive in dealing with the digital divide that may result in the lack of inclusion of marginalized individuals in quantum healthcare fields. Table 3 summarizes solutions and prospects in place that overcome existing

Table 3: Future Trends and Solutions for Privacy and Ethics in Quantum Healthcare.

Trend/Solution	Description	Implementation Timeline	Key Benefits	Primary Challenges
Quantum-Blockchain Integration	Hybrid systems combining quantum security with distributed ledgers	3–5 years	Immutable audit trails, quantum-resistant consensus	Scalability, interoperability
Post-Quantum Cryptography Standards	NIST-approved quantum-resistant algorithms for healthcare	1–3 years	Protection against quantum attacks	Migration complexity
Federated Quantum Learning	Distributed quantum ML preserving data locality	5–10 years	Privacy-preserving collaboration	Quantum network requirements
Room-Temperature Quantum Devices	Photonic and topological quantum processors	10–15 years	Reduced infrastructure requirements	Technical maturity
Global Quantum Health Initiative	International cooperation framework	3–5 years	Coordinated development	Political barriers

constraints, including implementation schedules, key advantages, and the main challenges of each technological and governance innovation.

8. Conclusion

The topic of combining quantum computing with smart hospitals and IoMT landscapes is a transformative force in the realm of health care that will grant incomparable benefits of medical progress whilst also creating significant issues regarding the violation of privacy and ethical regulations. This holistic review has indicated that the practical application of quantum healthcare technologies cannot be turned into a solely technical affair but must involve a radical reconsideration of how medical data privacy and patient rights are envisioned and established in quantum environments and the ethical considerations involved.

The evidence provided makes clear that the potential threat of quantum computers to current illegal encryption necessitates taking steps to begin to implement quantum-resistant encryption systems now, even though the timeframe is still open before the world can expect to face the threat of large-scale quantum computers. Healthcare companies are at a window of opportunity to harden sensitive medically related data against future quantum attacks and to have to spend a considerable amount of money on quantum-safe infrastructure, even though they have some budget issues temporarily. The review of quantum key distribution applications alongside post-quantum cryptography implementations proves that the technical existence of a solution does not solve the uncertainties of assimilating such systems in complicated healthcare systems due to the complexities of operations and their cost-effectiveness in an organizational set-up.

Ethical concerns of such nature go way beyond the straightforward issues of technical privacy to raise deeper implications to human dignity, autonomy, and justice in quantum-enabled medical interventions. The probabilistic nature of quantum computations is a problematic area that challenges traditional informed consent approaches, and since quantum algorithms may provide a means of emphasizing current health disparities by the process of biased pattern recognition, they may pose a hazard that could generate new forms of health disparity.

The proposed ethical-privacy framework of quantum IoMT is the implementation of an ordered methodology of integrated technical, organizational, and governance mechanisms. The multi-layered defense approach, dynamically managed consent, and close moral monitoring proposed in this framework provide a specific way forward to enable healthcare companies to benefit economically and ethically in the quantum computing age.

In the near future, evolving technologies like blockchain, AI, and advanced biotechnology will converge with quantum computing, therefore amplifying the level of privacy and ethical issues that require perpetuated adaptations in protective strategies and ethical paradigms. The advance of fault-tolerant quantum computers will exponentially increase both potentials and threats and necessitate constant restructuring of security precautions and ethics. To cross national borders and leave traditional spheres of jurisdiction of the regulatory authorities, quantum medical technology needs corresponding international cooperation in establishing quantum healthcare standards.

The real challenge is how to strike the right balance between technological progress and ethical protection mechanisms, and the only way of doing it is by applying adaptive governance rules that can keep pace with quantum capabilities. The key to building trust in quantum healthcare systems is the ability to show a commitment to privacy protection, ethical practice, and equitable access that transcends the mere technical ability to do so to organizational values and social responsibility.

Quantum healthcare, or even the holy goal behind quantum healthcare, should always be to make people healthier and happier with alcohol-free social justice and trustworthy innovation. At the present day, approaching the era of quantum healthcare, the decisions that will be made now will define whether quantum computing can become a human liberation and healing agent or another source of healthcare disparity in the delivery. It is left up to present leaders in healthcare, technology, and policy to create a base ensuring that quantum computing is used in the best interest of humanity and provides protection of the weak without devaluing the unique bond between patients and doctors.

References

Ahmed, S. F., Alam, M. S. B., Afrin, S., Rafa, S. J., Rafa, N., and Gandomi, A. H. (2023). Insights into Internet of Medical Things (IoMT): Data fusion, security issues and potential solutions. *Information Fusion*, 102: 102060. https://doi.org/10.1016/j.inffus.2023.102060

Banerjee, S., Barik, S., Das, D., and Ghosh, U. (2023). EHR Security and Privacy Aspects: A Systematic Review. 243–260. https://doi.org/10.1007/978-3-031-45878-1_17

Chapter 8

Healthcare IoT Architecture for Sensor-Driven Patient Care

Maranco Murugaiyan,[1,*] *M. Markco,*[2] *C. Murale*[3] and *R. Logeshwari*[1]

1. Introduction

Smart hospitals turn into complex ecosystems where different sets of sensors, medical equipment, and data analytics systems work together as systems to make sure that the patient and the clinic's support systems can be controlled in real time. The way these systems are built will determine how much real-time health information they can give and what safety and security standards must be met in medical environments.

Healthcare IoT architecture is a sophisticated integration of sensing, connectivity, data processing, and clinical applications into a system designed to enhance patient outcomes by continuous monitoring and measurement coupled with data-driven decision-making (Javaid et al. 2022). The system has many levels of technology gadgets that work together to filter physiological measurements, send information safely, assess medical information, and give medical professionals actionable intelligence. Because these systems are so complicated, interoperability standards, scalability needs, and regulatory compliance paradigms that govern how medical devices should be integrated should be carefully considered.

The transition from the difficult periodic care traditional delivery channel of monitoring individual health conditions to a continually connected sensor-driven

[1] Department of Networking and Communication, College of Engineering and Technology, SRM Institute of Science and Technology, Kattankulathur, Tamilnadu, India.
Email: loge.shwari54@gmail.com, https://orcid.org/0000-0002-6167-8996
[2] Department of Artificial Intelligence and Date Science, KIT - Kalaignarkarunanidhi Institute of Technology, Coimbatore-641 402 Tamil Nadu, India.
Email: markcome88@gmail.com, https://orcid.org/0009-0004-1851-8745
[3] Department of Information Technology, Coimbatore Institute of Technology, Coimbatore -641014, Tamilnadu, India.
Email: murale@cit.edu.in, https://orcid.org/0000-0001-8940-2142
* Corresponding author: maranco.m.phd@gmail.com, https://orcid.org/0000-0003-3965-360X

DOI: 10.1201/9781003783855-10

patient monitoring system has enabled the early detection of health decline, the optimization of treatment to ensure optimal patient care, and the implementation of preventive interventions. IoT-enhanced medical systems allow healthcare practitioners a level of visibility never before imaginable since they monitor vital signs in real time, keep track of medication compliance, assess environmental conditions, and keep an eye on behavioral trends. This kind of comprehensive data collection can help healthcare providers find potential problems before they become life-threatening, improve treatment plans based on how people respond to them, and keep treatment success going through more feedback loops.

Healthcare organizations develop new buildings with IoT technology that serves many goals at once, such as making patients safer, making clinical work more efficient, lowering costs, and making sure they follow the rules. As technology improves, sensor networks can now operate together with electronic health records, clinical decision support systems, and mobile healthcare to create strong, coordinated platforms that can help with care coordination across departments and specialties. The architecture should be able to handle different types of sensors, different communication demands, different processing needs, and different user interface needs, but it should also be able to support the same levels of performance and security.

To make patient care more efficient with sensors, various basic problems must be solved, such as data integration, system interoperability, data protection, and improving the flow of clinical work. The success is due to designer structures that find a compromise between technical innovation and practical issues like training staff, maintenance procedures, upgrades that happen at night, and long-term sustainability. The next parts talk about these architectural features and how they help improve patient care through smart hospital buildings.

2. Fundamentals of Healthcare IoT Architecture

Healthcare IoT is made up of many parts that work together to keep an eye on patients and coordinate care. These parts include physical devices, networking, data processing, and clinical apps. It finds the system's performance characteristics, such as how accurate the data is, how quickly it responds, the scaling thresholds, and how secure it is. Healthcare organizations use this information to help them make implementations that meet clinical needs, work well, and follow the rules. The architecture must take into account several functional areas, such as patient monitoring, clinical decision support, resource management, and administrative coordination. It should focus on issues related to system-wide complexity, performance and optimization, and maintenance processes.

2.1 Core Components of Healthcare IoT Systems

The architecture of Healthcare IoT systems has five layers, and each one can improve patient monitoring and clinical assistance (Ar-Reyouchi et al. 2022, Čolaković 2023). The perception layer is responsible for collecting patient data using physical sensors and medical devices, while the network layer manages data transmission utilizing various protocols. The edge computing layer enables real-time data processing and real-time reactions on severe patient state concerns. The cloud computing layer provides centralized data storage, state-of-the-art analysis, and integration with hospital information systems. At the application layer, there must be provided user interfaces and clinical decision support tools to support the administration of healthcare providers in delivering patient care. The sensor ecosystem incorporates wearable sensors to measure vital signs, environmental sensors used to track room conditions, medication dispensing sensors to track adherence, and equipment sensors used to detect the status of medical devices. Communication infrastructure plays a significant role in the reliability of the system, data security, and efficiency (Ma et al. 2022).

Healthcare IoT systems incorporate a variety of technologies, such as Wi-Fi networks when high bandwidth is desired, Bluetooth connections when a small personal area network is required, cellular networks when mobile devices are to be connected, and dedicated medical device networks when high-priority patient monitoring equipment is involved.

2.2 Layered Architecture Model

The healthcare IoT architecture is divided into layers, allowing for independent optimization of each layer while maintaining consistency in data flow and results. Because of this layered model, it is easier to keep things running and add new features. You can make changes without having to start over from scratch. There are tiers for sensors, perception, networks, edge computing, the cloud, and applications in the architecture stack. Data may flow both up and down between multiple levels and both up and down between multiple healthcare components, like patient sensors, hospital networks, edge analytics, cloud storage, and clinical apps. The architecture provides a stable and reliable performance and makes it easier to maintain without the need to redesign the system.

The entire architecture is shown in Figure 1 with sensors on the bottom and applications on the top. Data flows are two-way between layers, and layers have their own healthcare components, such as patient sensors, hospital networks, edge analytics, cloud storage, and clinical apps.

All healthcare Internet of Things systems use the perception layer, which includes physical sensors and medical equipment, such as data-gathering devices, in direct interaction with patients and healthcare facilities. This layer also makes use of wearable biosensors that permit continuous monitoring of vital signs, implantable sensors that permit the calibration of chronic illnesses, environmental sensors that measure air quality and temperatures, and medical equipment sensors that determine the state of the equipment and their functionality. It must be defined that the Perception Layer must have high accuracy parameters and be durable in a hostile environment like a hospital setting that consists of electromagnetic activity, temperature changes, and regular cleaning.

The sensors will transmit data to the processing systems through the network layer in various communication protocols that are most appropriate to various healthcare applications. Low-power networks provide battery-powered sensors that

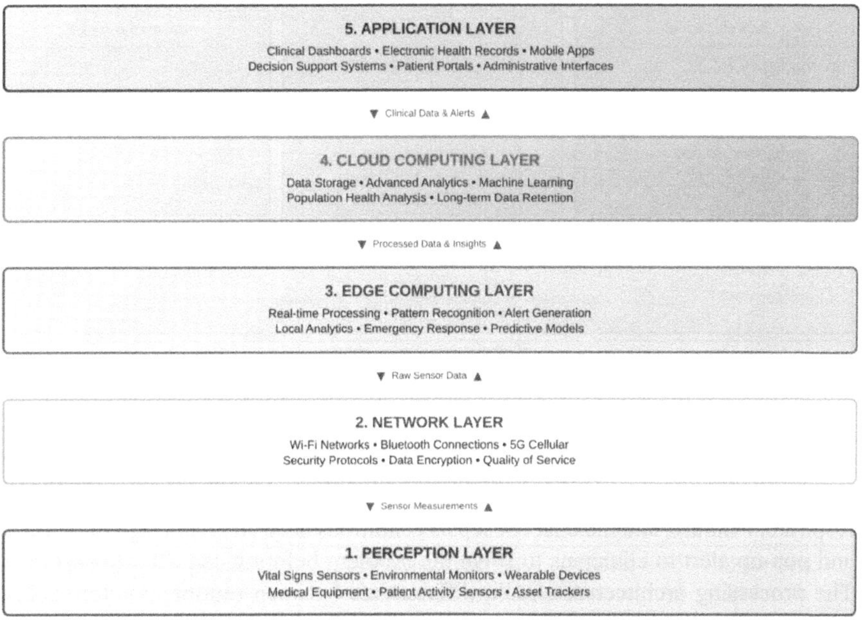

Figure 1: IoT Architecture in Healthcare Five-layer.

cannot afford to run for a short time, whereas high-bandwidth networks are better for medical imaging data and video surveillance systems. Security mechanisms that are implemented at the network layer include encryption, authentication, and access control. These techniques secure patient data while it is being sent, but they also keep communication reliable, which is important for patient monitoring applications.

The edge computing layer will be able to process data locally so that doctors can respond to emergency situations in a patient right away without having to connect to the cloud. The medical edge systems may look for trends and analyze data in real time to send alerts about important medical occurrences including heart attacks, breathing problems, and prescription mistakes. This layer reduces the amount of bandwidth the network needs and makes the system more responsive by letting local decision-making happen, which makes it easier to make quick clinical decisions.

2.3 Data Flow and Processing Architecture

Healthcare IoT systems create a lot of data all the time, so getting clinical insights from it depends on a processing architecture that can handle it without slowing down the system. Local data flow management consists of the steps of gathering, sending, storing, analyzing, and showing data. These steps need to work together without any problems so that patients and clinical decision-making may be monitored all the time. The architecture of the processing defines the capabilities of the system in recognizing patterns, detecting anomalies, predictive analytical tasks, and issues of such anomaly detection.

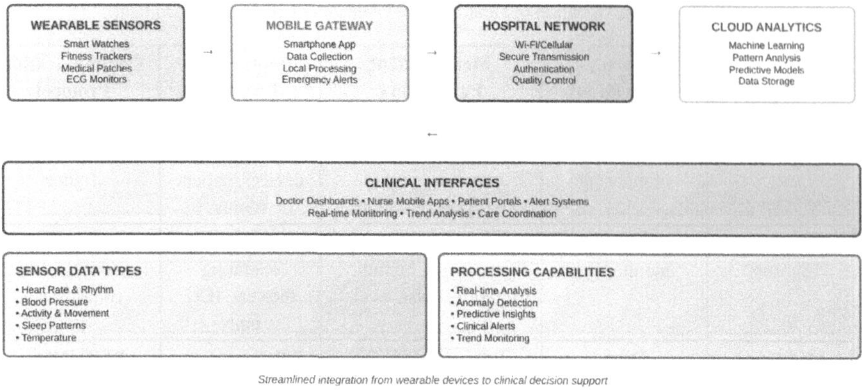

Figure 3: Wearable Sensor Integration Architecture.

and clinical performance. Lighting sensors keep track of how much light is in a room and help people with their circadian rhythms, which can change how fast they move, how they feel, and how well they sleep. Automated lighting control systems change the brightness and color temperature based on the time of day, the patient's preferences, and the needs of the clinic. Connectivity to patient monitoring systems allows custom lighting schedules that both follow a natural sleep pattern and allow adequate lighting to meet clinical needs.

Noise monitoring sensors measure sound levels that bother patients during rest time, stress, and in general their healing experience. Unnecessary noise in a medical setting has been associated with patient stress, impaired healing, and exhaustion in medical professionals at the expense of care quality. Noise monitoring in real time allows programmed alerts in cases where the sound reaches beyond the recommended limits and also supports policies that would keep therapeutic conditions favorable to patient healing. Figure 3 illustrates the end-to-end process of data acquisition via wearable sensors, smartphone gateways, hospital Wi-Fi networks, processing equipment, cloud-utility solutions, and clinical monitor dashboards to show data movement, computation, and decision points.

4. Network Infrastructure and Connectivity

Healthcare IoT systems require a robust network infrastructure that provides secure, reliable, and high-performance connections between sensor networks and centralized processing systems. This structure should support real-time patient monitoring, data gathering, device mobility, and external healthcare system bonding. Network architecture choices significantly impact system performance, scalability, security, and maintenance needs. Healthcare network environments are complex, requiring specialized planning to handle multiple tasks simultaneously and maintain reliability for patient safety. Critical patient monitoring systems require high bandwidth, minimal latency, and efficient data transfer without crashing essential systems. Network architecture should prioritize traffic while maintaining system performance and reliability.

4.1 Wireless Communication Protocols

Healthcare IoT systems also apply various wireless protocols that are specific to different types of sensors, data transmission demands, and environments. The process of selecting proper communication protocols determines the quality of sensor battery life, reliability of data transmissions, the amount of bandwidth used by the network, and the efficiency of the security configuration. Knowledge of the characteristics of the protocols can help to make architectural choices that can deliver improved system performance without undermining compatibility with the available hospital information technology infrastructure.

Wi-Fi networks have high bandwidth and could be used to communicate with medical devices that generate large data volumes such as medical equipment, video surveillance, and advanced patient monitoring systems (Yew et al. 2020). Enterprise-level security measures on hospital Wi-Fi networks include network segmentation, which separates the traffic of medical devices to a standalone segment and isolates it so that devices stay connected to the clinical information systems. The quality-of-service tweaking will offer priority bandwidth to essential patient monitoring programs whilst offering normal administrative network traffic.

Bluetooth Low Energy technologies make communication with wearable sensors and personal health monitoring devices with a long battery power supply and low power consumption possible and within a range of proximity distance. The low-power dissipation feature enables the long-term usage of the application in patient monitoring, whereas the data transfer is reliable during routine health examinations. Bluetooth mesh networking technologies permit extending to terrestrial areas of the patient with the sensor network without losing central connectivity in order to collect and analyze the data.

Zigbee-based protocols offer mesh networking features that allow deployment of distributed sensors throughout a healthcare facility and still keep power requirements low, even in a battery-powered sensor unit. The mesh topology serves to add redundancy to the networks so that the network itself does not go down even when individual sensors or network nodes fail. Given that Zigbee allows connecting hundreds of devices, it can be used to extensively monitor the environment and locate assets across the hospital setting.

Cellular communication, including 5G networks, supports the use of ambulatory care, remote patient monitoring, and emergency response applications that necessitate wide-area network coverage (Devi et al. 2023, Ahad et al. 2023). Real-time telemedicine applications and remote surgery are based on real-time requirements to transmit and receive data and are supported by the low-latency characteristics of 5G networks. As it applies to telemedicine, the network slicing feature enables a healthcare organization to create specific segments in its network with assured performance quality to ensure that its life-saving applications perform as expected.

4.2 Edge Computing Integration

The integration of edge computing in a healthcare IoT architecture allows computation to be done locally, resolving time-sensitive patient conditions in real

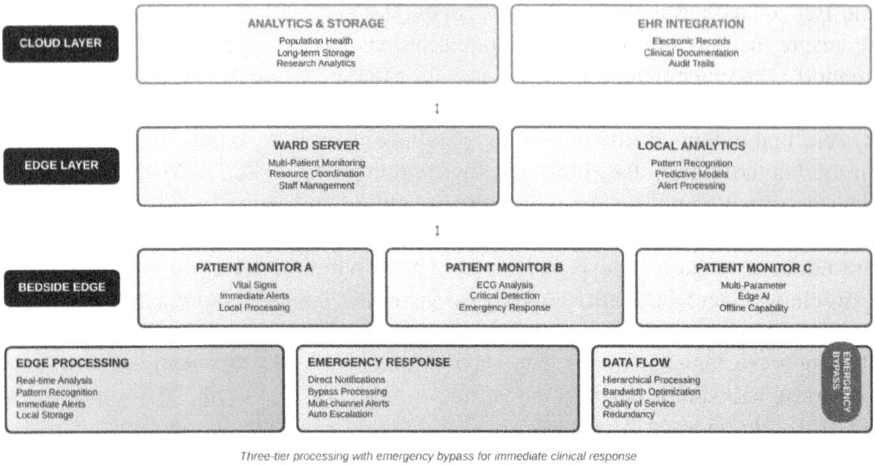

Figure 4: Edge Computing Architecture for Real-Time Patient Monitoring.

time and minimizing bandwidth while increasing system reliability. Edge systems analyze sensor data on the edge, determine emergency situations, and perform real-time analytics to generate clinical alerts that are independent of cloud connectivity. Efficient patient safety relies on this local processing capability, as network communication delays and other network connectivity disruptions are unacceptable.

Healthcare edge computing equipment provides dedicated processing algorithms adapted to the analysis of medical data including aberration recognition, abnormal condition detection, medication fault analysis, and critical condition recognition. Figure 4 illustrates a hierarchical processing structure where sensor data are collected and analyzed at the bedside by edge devices, the information is collected on intermediate edge servers and processed on cloud-based systems, offering comprehensive analytics and long-term storage, with emergency alert pathways presenting potential bypassing normal processing to generate alerts to clinicians.

4.3 Cloud Integration and Data Management

The healthcare business environment is in need of architecture design to strike the optimal balance between the benefits of the cloud and the requirements for security, regulatory compliance, and data sovereignty.

Healthcare cloud environments provide particular data management features such as encrypting medical data, keeping track of audit trails, making backups and recovering from disasters, and generating reports for regulatory compliance. These platforms require the capability to comply with the requirements of the Healthcare Insurance Portability and Accountability Act, continuing to support high levels of availability and performance required by clinical operations. The multi-tenant cloud design is capable of facilitating resource sharing with the assurance of data isolation and data privacy for various healthcare bodies and patient groups. Figure 5 presents the full cloud integration ecosystem that supports data ingestion into several hospital

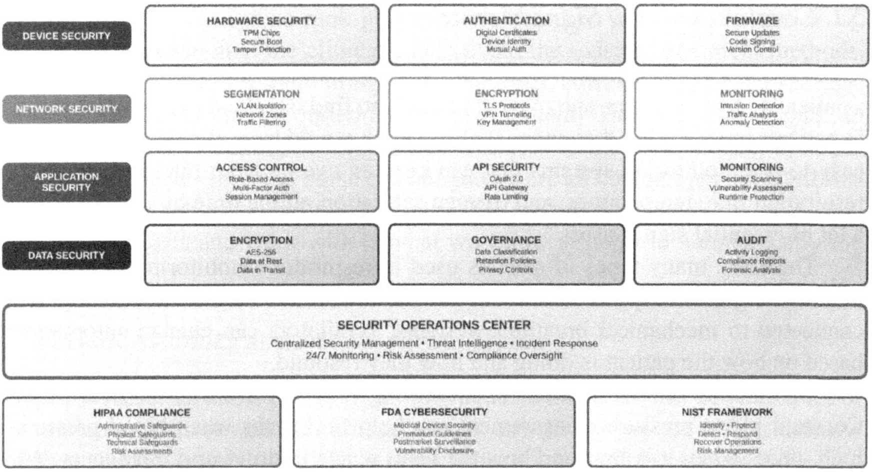

Figure 5: Healthcare IoT Cloud Integration Architecture.

edge systems, a cloud-based analytics platform, electronic health records integration, connections to third-party systems like laboratory results and pharmacy systems, and clinical dashboard delivery to various device types—desktop workstations, mobile devices, and tablet computers.

Cloud systems offer advanced data analytics tools for population health analysis, treatment prediction, clinical research, and operation optimization. These capabilities utilize machine learning systems to make evidence-based treatment plans, evaluate medication effectiveness, and measure patient risk stratification. Cloud platforms provide enormous computing power for these capabilities. Interoperability with third-party healthcare systems requires secure data exchange methods, ensuring patient privacy and care coordination. Health Level Seven International standards and Fast Healthcare Interoperability Resources protocols facilitate standardized data transfer between health organizations and technology platforms, ensuring continuity of healthcare across multiple providers.

5. Real-Time Patient Monitoring Systems

IoT architecture in healthcare is crucial for real-time patient monitoring, enabling early intervention and better clinical outcomes. These systems combine sensor data, algorithms, and clinical user interfaces to provide extensive monitoring beyond traditional episodic systems. They ensure immediate alerts for critical patients, reduce false alarms, and ensure system reliability. Modern patient monitoring devices process multiple physiological variables to provide a comprehensive picture of patient conditions and detect discomfort. These systems let people create tailored care plans and raise the level of care based on evidence, but they need to be aware of clinically important changes and be detailed so that users do not get tired of being alarmed.

6. Data Analytics and Clinical Decision Support

Healthcare IoT devices produce enormous volumes of data, requiring analytics platforms to derive clinical insights and facilitate evidence-based decision-making. Healthcare practitioners can use real-time data processing and advanced analytics to figure out what will happen to a patient and how to treat them. But these systems have to find a balance between how complicated the processing is and how fast it is in real time, as well as how accurate it is in the clinic and how well it follows the guidelines. Advanced analytics platforms use machine learning algorithms that are based on data from previous patients to make predictions about real-life circumstances. These systems must strike a balance between the individual demands of the patient and adherence to clinical guidelines and evidence-based treatment protocols.

6.1 Real-Time Data Processing and Analysis

Healthcare IoT systems may need to process data in real time, so they need complicated algorithms that can quickly and consistently handle different data streams without causing delays that could make life-saving apps less responsive. Varying types of sensors should be able to send data at varying speeds to processing systems. There should also be quality checks in place to make sure that the clinical accuracy and reliability are high. It should be able to conduct a lot of analysis because it demands a lot of computing power for complicated forms of study. In systems that use continuous sensor data, platforms process the data in real time to uncover crucial clinical events, like changes in vital signs, how well medications work, and how well therapeutic treatments work. The platforms use sliding window methods to look at the trends in recent data while still maintaining the historical context that is needed for appropriate clinical assessment. Real-time processing makes it possible to quickly find potential problems such as cardiac arrhythmias, respiratory distress, and neurological abnormalities that are serious and need rapid medical attention (Muralitharan et al. 2021).

Pattern recognition algorithms use sensor data to find clinical problems that may get worse over time, such as sepsis, worsening heart disease, and side effects of drugs. Such algorithms must be in a position to distinguish between normal physiological changes and pathological changes. They must also consider factors such as the age of a patient, medical history, treatment plan, and physiological baseline. More powerful pattern recognition systems apply ensemble techniques. These strategies combine methods so that one or more forms of analysis are used to make diagnoses more precise and reduce false positives.

To identify an anomaly in patient information, anomaly detection systems are always reviewing patient information to identify unusual patterns that might indicate the possibility of a sensor, equipment, or new clinical condition issue. These systems must be capable of distinguishing between sensor artifacts and actual physiologic changes in the body, as well as be conscious of significant clinical changes. Machine learning has adaptive anomaly detection. This implies that you compare a model of normal activity with each patient individually, but will still detect a significant deviation due to therapeutic causes.

6.2 Predictive Analytics for Patient Outcomes

It is reasonable to conclude that predictive analytics can be used to examine both past and current data on patients, make predictions about their clinical outcomes, alleviate the risk of complications, and make the treatment process more efficient depending on the predicted effects.

Early warning systems make use of real-time data on patient monitoring to make predictive assessments of clinical deterioration even before the emergence of overt symptoms, and therefore enable pre-emptive measures to prevent adverse outcomes. The scoring systems apply an algorithm to determine the riskiness of a patient by using study outcomes and vital signs, lab tests, and clinical observations regarding this patient along with his or her risk factors. It has been demonstrated that early warning systems provide a significant improvement in patient outcomes since sepsis, heart attacks, and respiratory arrest are noted sooner (Shamout et al. 2020). Medication response prediction applies patient monitoring data to predict treatment efficacy and to anticipate adverse drug reactions prior to their onset as they become clinically visible.

6.3 Machine Learning Applications in Patient Care

The models are built on patient data, training data sets, clinical evaluations, and known outcomes. This necessitates extensive testing to ensure both the generalizability and clinical dependability of the validation. Unsupervised learning approaches simply show the hidden patterns in patient data. They can help us learn about clinical phenomena we did not know about before or tell us how well therapy works. The clustering algorithms group patients with similar traits into smaller groups so that they can get more specialized care or be watched more closely. Deep learning algorithms analyze sensor data to identify intricate patterns applicable in clinical diagnosis, thereby enhancing treatment. These kinds of systems consume a lot of resources and need to be evaluated and confirmed regularly to make sure they are always accurate, valid, and effective in changing clinical settings.

7. Security and Privacy Considerations

IoT solutions in healthcare must unequivocally safeguard patient data. This means they need to check the devices, encrypt the data, protect the networks, and control who may access the data. These measures must satisfy privacy standards and make sure that the system operates and that clinical work is done quickly. Security considerations must address technical vulnerabilities and operational threats that may harm patient data or system reliability. The distributed sensor networks in IoT require organized security measures across all system elements, supporting data at rest, in transit, and during processing. The architecture must balance security concerns with operational efficiency to prevent disruptions in clinical processes and patient care delivery.

7.1 Data Encryption and Secure Communication

Data encryption of healthcare IoT systems prevents the leakage of patient data over the air and in storage and keeps the data safe within the rules of the healthcare privacy data protection, such as HIPAA and international regulations (Almalawi et al. 2023, Das and Namasudra 2022). When utilizing encryption protocols to make sure that real-time systems like life-saving patient monitoring work, the security strength vs. computational speed is a big problem. The architectural design should use end-to-end encryption to protect data all the way through the processing pipeline, from when the sensor collects it to when it is shown to the doctor.

Transport layer security mechanisms stop data in transit between IoT sensors and processing systems, but they do let you validate the integrity of data and authenticate it. Advanced encryption standards make sure that only authorized users can read intercepted data, and digital signatures can prove that data is correct and show that someone tried to change it.

Secured communication algorithms make sure that only the proper person can see the sensor data and that no one else can change it. Message integrity verification checks to make sure that no changes have been made to the clinical data while it is being sent, and non-repudiation methods keep track of everything to make sure that the rules are being followed.

End-to-end encryption protects the data stored on IoT sensors and edge computing devices and keeps the keys safe. However, if the physical IoT sensors and edge computing devices are hacked, the data will be lost. Hardware security modules provide cryptographic processing and tamper-resistant key storage, which makes the device more secure without compromising how well it works. Strong secure firmware-update mechanisms let security patches be added without putting the device's operation or patient monitoring at risk.

7.2 Access Control and Authentication Systems

In the healthcare IoT architecture, access control systems provide role-based access controls that only allow authorized users to access patient information and controls to the extent they need to in their clinical work. Authentication mechanisms have to address the requirement of operational efficiency with security standards to avoid security problems at the operational level and meet efficiency needs in time-critical situations. Systems with multi-factor authentication provide an increased level of security in addition to making it easy to use in stressful healthcare settings.

Biometric authentication systems offer secure controlled access, in addition to the quick clinical authentication that is required during emergencies when a password-based authentication system may be slow in granting access to vital patients. Fingerprinting devices, facial recognition, and voice verification allow the use of patient monitoring systems by healthcare providers in a fast and secure way. Capability to integrate with hospital badge systems allows authorization using proximity that automatically logs healthcare providers into the right systems depending on their location and assigned roles.

User access duration control—user access duration is controlled by user access management systems that time out active sessions and disconnect idle connections to prevent unattended user access to patient data via unattended workstations or mobile devices. Such systems must allow security requirements to meet clinical workflow requirements that might demand long periods of access when treating a patient. Adaptive session management capabilities change the timeout durations based on the user's role, location, and current clinical activity to provide the right level of security at the same time.

Audit logging systems keep account of all attempts to access, query data, and change the system. This makes it easy to audit device use and probable security breaches with healthcare IoT systems. These logs should record enough information to allow forensics but not to the degree that violates patient privacy or fails to allow regulatory compliance. Automated audit analysis detects irregular access patterns that may signal security violations or non-compliant data access and assists in compliance reporting efforts.

7.3 Compliance with Healthcare Privacy Regulations

Healthcare IoT systems must comply with various privacy laws, including the Health Insurance Portability and Accountability Act, the European General Data Protection Regulation, and local healthcare laws. These laws regulate patient data gathering, processing, storing, and sharing, and affect architectural design choices such as data storage locations, processing methods, access controls, and data transfer protocols. HIPAA compliance requires administrative, physical, and technical safeguarding to secure patient health information throughout the system's life. Technical safeguards include access controls, audit logging, encryption, and log-off mechanisms, while physical safeguards include secure facility access/entry, workstation controls, and media disposal processes. Data minimization requires healthcare IoT systems to gather, analyze, and retain minimal patient information for clinical applications.

Data retention policies automate the elimination of expired information, balancing clinical requirements with privacy. Regulated automated data lifecycle management ensures compliance with retention requirements and supports clinical research initiatives. Global data transfer regulation is crucial for cloud computing design and neuro-technology companies designing healthcare IoT systems.

8. Implementation Challenges and Solutions

Implementing IoT in healthcare comes with its own set of problems, such as interoperability, integration complexity, and organizational factors that can affect the success of certain systems. There are many different technological platforms, communication standards, data formats, and vendor platforms that need to work together with as little effect on clinical workflow efficiency and compliance with regulations as possible.

Interoperability is hard to achieve because the healthcare setting has a mix of different technologies, such as electronic health records, laboratory information systems, pharmacy management software, and medical device networks. These

The transition to intelligent healthcare IoT systems utilizing artificial intelligence, advanced analytics, and innovative communication methods may enhance clinical capacities in the future. This would create new ways to offer personalized medicine, predictive health, and health management, which would make the population healthier. Healthcare facilities need to invest in the IoT for a long time to get the most out of their money.

The healthcare IoT infrastructure is all about using smart technology to keep an eye on things in real time, get insights from data, and take action to improve health and clinical outcomes. To make these systems better, clinical stakeholders, technology experts, and regulatory experts need to work together to find a balance between what the technology can do and what the clinical needs are, like patient safety, data security, and the long-term viability of operations.

References

Ahad, A., Tahir, M., Aman Sheikh, M., Ahmed, K. I., Mughees, A., and Numani, A. (2020). Technologies Trend towards 5G network for smart health-care using IoT: A review. *Sensors* (Basel, Switzerland), 20(14): 4047. https://doi.org/10.3390/s20144047

Ahad, A., Ali, Z., Mateen, A., Tahir, M., Hannan, A., Garcia, N. M. et al. (2023). A Comprehensive review on 5G-based Smart Healthcare Network Security: Taxonomy, Issues, Solutions and Future research directions. *Array*, 18: 100290. https://doi.org/10.1016/j.array.2023.100290

Ali, Z., Ghani, A., Khan, I., Chaudhry, S. A., Islam, S. H., and Giri, D. (2020). A robust authentication and access control protocol for securing wireless healthcare sensor networks. *Journal of Information Security and Applications*, 52: 102502. https://doi.org/10.1016/j.jisa.2020.102502

Alladi, T., Naren, N., and Chamola, V. (2020). HARCI: A two-way authentication protocol for three entity healthcare IoT networks. *IEEE Journal on Selected Areas in Communications*, 39(2): 361–369. https://doi.org/10.1109/jsac.2020.3020605

Almalawi, A., Khan, A. I., Alsolami, F., Abushark, Y. B., and Alfakeeh, A. S. (2023). Managing security of healthcare data for a modern healthcare system. *Sensors* (Basel, Switzerland), 23(7): 3612. https://doi.org/10.3390/s23073612

Alshamrani, M. (2021). IoT and artificial intelligence implementations for remote healthcare monitoring systems: A survey. *Journal of King Saud University - Computer and Information Sciences*, 34(8): 4687–4701. https://doi.org/10.1016/j.jksuci.2021.06.005

Ar-Reyouchi, E. M., Yahiaoui, R., Rattal, S., Ghoumid, K., Ar-Reyouchi, D., and Elmazria, O. (2022). Protocol wireless medical sensor networks in IoT for the efficiency of healthcare. *IEEE Internet of Things Journal*, 9(13): 10693–10704. https://doi.org/10.1109/jiot.2021.3125886

Clay, I., Angelopoulos, C., Bailey, A. L., Blocker, A., Carini, S., Carvajal, R. et al. (2021). Sensor data integration: A new cross-industry collaboration to articulate value, define needs, and advance a framework for best practices. *J. Med. Internet Res.*, 23(11): e34493. https://doi.org/10.2196/34493

Coffey, J. D., Christopherson, L. A., Williams, R. D., Gathje, S. R., Bell, S. J., Pahl, D. F. et al. (2022). Development and implementation of a nurse-based remote patient monitoring program for ambulatory disease management. *Front. Digit. Health*, 4: 1052408. doi: 10.3389/fdgth.2022.1052408

Čolaković, A. (2023). IoT systems modeling and performance evaluation. *Computer Science Review*, 50: 100598. https://doi.org/10.1016/j.cosrev.2023.100598

Daryanani, A. E., Maduekwe, U. N., Baird, P., and Ehrenfeld, J. M. (2025). Ensuring medical device safety: The role of standards organizations and regulatory bodies. *Journal of Medical Systems*, 49(1). https://doi.org/10.1007/s10916-025-02150-x

Das, S., and Namasudra, S. (2022). A novel hybrid encryption method to secure healthcare data in IoT-enabled healthcare infrastructure. *Computers and Electrical Engineering*, 101: 107991. https://doi.org/10.1016/j.compeleceng.2022.107991

Devi, D. H., Duraisamy, K., Armghan, A., Alsharari, M., Aliqab, K., Sorathiya, V. et al. (2023). 5G Technology in Healthcare and Wearable Devices: A Review. *Sensors*, 23(5): 2519. https://doi.org/10.3390/s23052519

Hartmann, M., Imran, A., and Hashmi, U. S. (2019). Edge computing in smart health care systems: Review, challenges, and research directions. *Transactions on Emerging Telecommunications Technologies*, 33(3). https://doi.org/10.1002/ett.3710

He, L., Eastburn, M., Smirk, J., and Zhao, H. (2023). Smart chemical sensor and biosensor networks for healthcare 4.0. *Sensors*, 23(12): 5754. https://doi.org/10.3390/s23125754

Javaid, S., Zeadally, S., Fahim, H., and He, B. (2022). Medical sensors and their integration in wireless body area networks for pervasive healthcare delivery: A review. *IEEE Sensors Journal*, 22(5), 3860–3877. https://doi.org/10.1109/jsen.2022.3141064

Kanungo, P. (2025). Edge computing in healthcare: Real-time patient monitoring systems. *World Journal of Advanced Engineering Technology and Sciences*, 15(1): 001–009. https://doi.org/10.30574/wjaets.2025.15.1.0168

Khan, N.-E., and Rudman, R. (2025). IoT medical device risks: Data security, privacy, confidentiality and compliance with HIPAA and COBIT 2019. *South African Journal of Business Management*, 56(1). https://doi.org/10.4102/sajbm.v56i1.4796

Ko, H. Y. K., Tripathi, N. K., Mozumder, C., Pal, I., and Muengtaweepongsa, S. (2023). Real-time remote patient monitoring and alarming system for noncommunicable lifestyle diseases. *International Journal of Telemedicine and Applications*, 2023, 1–13. https://doi.org/10.1155/2023/9965226

Lakshman, P., Gopal, P. T., and Khurdi, S. (2025). Effectiveness of remote patient monitoring equipped with an early warning system in tertiary care hospital wards: retrospective cohort study. *Journal of Medical Internet Research*, 27: e56463. https://doi.org/10.2196/56463

Leenen, J. P. L., Van Westreenen, H. L., Schoonhoven, L., Patijn, G. A., Van Dijk, J. D., and Leerentveld, C. (2020). Current evidence for continuous vital signs monitoring by wearable wireless devices in hospitalized adults: Systematic review. *Journal of Medical Internet Research*, 22(6): e18636. https://doi.org/10.2196/18636

Lin, X., Luo, J., Liao, M., Su, Y., Lv, M., Li, Q. et al. (2022). Wearable Sensor-Based Monitoring of Environmental Exposures and the Associated Health Effects: A Review. *Biosensors*, 12(12): 1131. https://doi.org/10.3390/bios12121131

Ma, S.-C., Alkhaleefah, M., Chang, Y.-L., Chuah, J. H., Chang, W.-Y., Ku, C.-S. et al. (2022). Inter-Multilevel Super-Orthogonal Space–Time Coding Scheme for Reliable ZigBee-Based IoMT Communications. *Sensors*, 22(7): 2695. https://doi.org/10.3390/s22072695

Muralitharan, S., Nelson, W., Di, S., McGillion, M., Devereaux, P., Barr, N. G. et al. (2021). Machine learning–based early warning systems for clinical deterioration: systematic scoping review. *J. Med. Internet Res.*, 23(2): e25187. https://doi.org/10.2196/25187

Nwanosike, E. M., Conway, B. R., Merchant, H. A., and Hasan, S. S. (2021). Potential applications and performance of machine learning techniques and algorithms in clinical practice: A systematic review. *International Journal of Medical Informatics*, 159: 104679. https://doi.org/10.1016/j.ijmedinf.2021.104679

Ravizza, A., De Maria, C., Di Pietro, L., Sternini, F., Audenino, A. L., and Bignardi, C. (2019). Comprehensive review on current and future regulatory requirements on wearable sensors in preclinical and clinical testing. *Frontiers in Bioengineering and Biotechnology*, 7(313). https://doi.org/10.3389/fbioe.2019.00313

Saleh, N., Haidar, A. M., and Kassem, A. (2018). Energy-efficient architecture for wireless sensor networks in healthcare applications. *IEEE Access*, 6: 6478–6486. https://doi.org/10.1109/access.2018.2789918

Selvaraju, V., Spicher, N., Wang, J., Ganapathy, N., Warnecke, J. M., Leonhardt, S. et al. (2022). Continuous monitoring of vital signs using cameras: A systematic review. *Sensors*, 22(11): 4097. https://doi.org/10.3390/s22114097

Shah, P. K., Irizarry, J., and O'Neill, S. (2018). Strategies for managing smart pump alarm and alert fatigue: A narrative review. Pharmacotherapy: *The Journal of Human Pharmacology and Drug Therapy*, 38(8): 842–850. https://doi.org/10.1002/phar.2153

Shamout, F., Zhu, T., and Clifton, D. A. (2020). Machine learning for clinical outcome prediction. *IEEE Reviews in Biomedical Engineering*, 14: 116–126. https://doi.org/10.1109/rbme.2020.3007816

Shoaib, M., Mustafee, N., Madan, K., and Ramamohan, V. (2023). Leveraging multi-tier healthcare facility network simulations for capacity planning in a pandemic. *Socio-Economic Planning Sciences*, 88: 101660. https://doi.org/10.1016/j.seps.2023.101660

Shruti, and Trivedi, N. (2023). Predictive analytics in healthcare using machine learning. 2023 14th *International Conference on Computing Communication and Networking Technologies (ICCCNT)*. https://doi.org/10.1109/ICCCNT56998.2023.10306782

Sidey-Gibbons, J. A. M., and Sidey-Gibbons, C. J. (2019). Machine learning in medicine: a practical introduction. *BMC Medical Research Methodology*, 19(1). https://doi.org/10.1186/s12874-019-0681-4

Sun, L., Ren, H., Jiang, X., and Guo, Y. (2020). Edge-cloud computing and artificial intelligence in internet of medical things: architecture, technology and application. IEEE Access, 8, 101079–101092. https://doi.org/10.1109/access.2020.2997831

Van Calster, B., Timmerman, D., Wynants, L., Collins, G. S., and Steyerberg, E. W. (2019). Predictive analytics in health care: how can we know it works? *Journal of the American Medical Informatics Association*, 26(12): 1651–1654. https://doi.org/10.1093/jamia/ocz130

Winters, B. D., Cvach, M. M., Bonafide, C. P., Hu, X., Konkani, A., O'Connor, M. F. et al. (2018). Technological distractions (Part 2): A summary of approaches to manage clinical alarms with intent to reduce alarm fatigue. *Critical Care Medicine*, 46(1). https://doi.org/10.1097/CCM.0000000000002803

Yang, Y., and Deng, Z. D. (2019). Stretchable sensors for environmental monitoring. *Applied Physics Reviews*, 6(1): 011309. https://doi.org/10.1063/1.5085013

Yaqoob, T., Abbas, H., and Shafqat, N. (2019). Integrated security, safety, and privacy risk assessment framework for medical devices. *IEEE Journal of Biomedical and Health Informatics*, 24(6): 1752–1761. https://doi.org/10.1109/jbhi.2019.2952906

Yew, H. T., Ng, M. F., Ping, S. Z., Chung, S. K., Chekima, A., and Dargham, J. A. (2020). IoT based real-time remote patient monitoring system. *2020 16th IEEE International Colloquium on Signal Processing & Its Applications (CSPA)*, 176–179. https://doi.org/10.1109/cspa48992.2020.9068699

Zhang, S., Cui, Y., Xu, D., and Lin, Y. (2025). A collaborative inference strategy for medical image diagnosis in mobile edge computing environment. *PeerJ Computer Science*, 11: e2708. https://doi.org/10.7717/peerj-cs.2708

Section 3
Applications and Use Cases

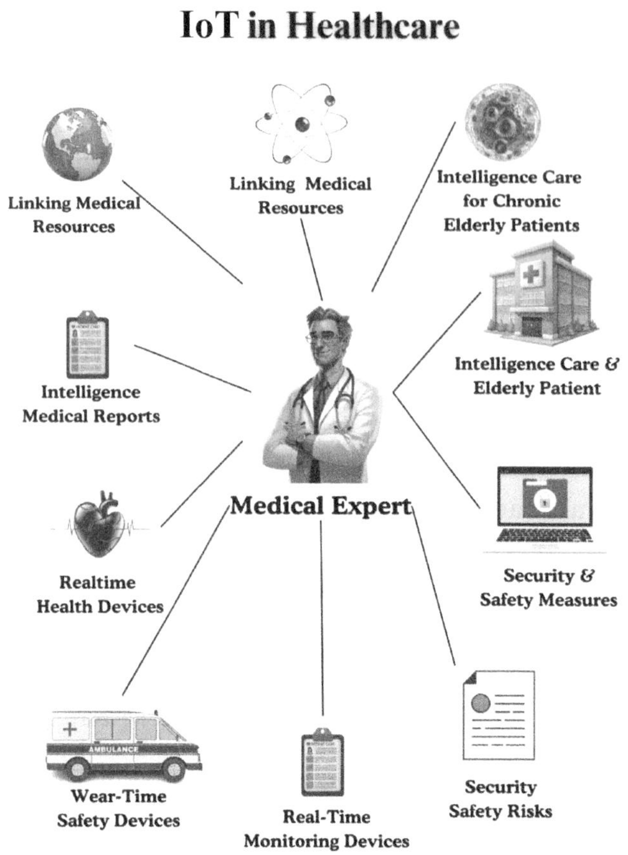

Figure 1: An Impression of Healthcare's Use of the IoT.

IoT has a positive effect on healthcare. AI applications are transforming these systems in a number of ways (Qadri et al. 2020). By enabling body-worn sensors to gather massive amounts of health data and encouraging cost savings, modernity, and personalization, the IoT helps healthcare (Habibzadeh et al. 2019). IoT enhances healthcare by connecting physical devices to the internet, providing real-time data, monitoring, and cost-effective solutions for elderly and chronic patients (Bhatt and Bhatt 2017). IoT enhances government financing, patient well-being, treatment quality, productivity, and decision-making through more complex and intelligent systems (Aghdam et al. 2021). With the IoT network, which consists of linked devices, sensors, and processors, healthcare apps may help patients (Laplante and Laplante 2016). IoT technology in healthcare uses billions of sensors for remote and digital patient monitoring, addressing global health issues like COVID-19 (Al-Kahtani et al. 2022). In order to handle healthcare issues including resource scarcity, population aging, and growing medical expenditures, the IoT is an essential system that combines hardware, software, and tangible objects (Ahmadi et al. 2019). IoMT, which provides remote care, elder care, and epidemic identification, was developed as a result of the healthcare industry's use of IoT to enhance services and remotely monitor patient vitals (Naresh et al. 2020). Patient safety, employee happiness, and healthcare operational efficiency have all improved as a result of the IoT's revolution in data transmission, which has raised demand for data analysis (Dang et al. 2019). The worldwide health business has been greatly impacted by the emergence of wearable health monitoring devices, which offer real-time data for physical examinations and smooth patient–doctor communication (Din et al. 2019).

Patient health and equipment efficiency have improved thanks to the Internet of Things, which is revolutionizing the collection of data and transmission in a wide variety of industries, including healthcare (Yeole and Kalbande 2016). Cost-cutting medical solutions, assisted living, high-quality infrastructure, smart devices, and connected resources are some of the key areas where IoT is playing a major role (Karthick and Pankajavalli 2020). Quality monitoring, cost optimization, and operational efficiency have improved thanks to IoT, which generates data and services for smart healthcare by connecting physical objects (Rghioui and Oumnad 2018). The IoT, a rapidly evolving technology, poses significant health and safety risks, especially in medical devices like wearables and implanted ones, due to system security issues (Somasundaram and Thirugnanam 2021). IoT can aid in managing equipment, inventories, and patient tracking, addressing issues like overstocking during emergencies, focusing on industry solutions and technology in healthcare (Lalitha et al. 2021). The growth of global healthcare infrastructure is made possible by IoT and WSNs, which overcome the high development and operating costs of commercial systems by giving stakeholders access to real-time information (Maksimović et al. 2015). The rapid evolution of technology necessitates a sophisticated smart healthcare framework, utilizing edge computing, 5G, and IoT sensors for real-time solutions and security (Amin and Hossain 2020). Healthcare IoT leverages IoT for e-health, supported living, and human-centric sensing, necessitating a multi-agent strategy using machine learning for predictive analytics to protect it from cybersecurity risks (MacDermott et al. 2019).

3. Machine Learning (ML) and Deep Learning (DL)

ML uses its environment to collect data (El Naqa and Murphy 2015). The data, along with new learning theories and algorithms, have all influenced recent advances in machine learning (Jordan and Mitchell 2015). Machine learning's primary benefit is its ability to operate autonomously after the users decide how to use the data (Mahesh 2020). Common types of ML algorithms have been described already (Ayodele 2010). ML is the primary focus of AI research (Wang and Zhou 2009).

The four main subfields of machine learning include learning complicated stochastic models, scaling SV learning algorithms, reinforcement learning, and classifiers (Dietterich 1997). Through talks and actions, ML enables computers to simulate human behavior and provide learning resources for later use (Alzubi et al. 2018). ML is useful for computer engineering. It produces fundamental statistical computational theories (Mitchell 2006). ML uses observed examples to extract general principles without specific instructions (Bzdok et al. 2017). Machine learning is a highly automated process that minimizes the need for human interaction (Tarca et al. 2007).

Deep Learning (DL) approaches in computational models enhance many applications (LeCun et al. 2015). DL is one of the ML methodologies that helps models learn aspects from unorganized input at various levels, and it outperforms already available methods (Mathew et al. 2021). There are many uses of DL (Deng and Yu 2014). Artificial neural networks (ANNs) are useful because of their capacity to handle large volumes of data and processing ability (Hao et al. 2016). DL is used more in academic research than ML (Du et al. 2016). DL generates multiple representations of the data distribution (Lauzon 2012).

DL is more advantageous than support vector machines (SVM) (Arnold et al. 2011). ANNs match input, output, nodes, and interconnections to facilitate data interchange and calculation (Kim 2022). Researchers are interested in DL because it enhances computer systems' prediction capabilities with better algorithms and large amounts of data. Applications for machine learning include object detection, self-driving automobiles, picture recognition, medication formulation, and illness diagnosis (Khan et al. 2019). Although DL, a well-liked AI method, is good at analyzing and classifying biological data, those who are not familiar with it may find it difficult to use (Webb 2018).

4. Machine Learning in Stroke Treatment

By reliably forecasting outcomes following endovascular therapy and resolving intricate interactions and non-linear correlations between factors, machine learning might enhance stroke care (Van et al. 2018) (Figure 2).

A machine learning method called Random Forest has demonstrated encouraging results in improving the outcomes of ischemic stroke treatment, offering vital information to healthcare providers (Khorsand et al. 2024). To guarantee compliance with clinical practice standards and collaborative decision-making in the selection of candidates and assessment of patient prognosis, hybrid machine learning techniques may be applied as a customized risk prediction model (Nie et al. 2024). ML can forecast stroke (Feyen et al. 2022). When it came to properly forecasting MCA strokes, DWI-FLAIR mismatch, and the effectiveness of endovascular therapy, ML methods demonstrated higher sensitivity (Gupta 2023). AI studies show conflicting results, suggesting overestimation of its genuine effect (Mahmoud et al. 2023). By identifying high-risk patients, guiding treatment options, and adding fresh data from extensive multi-institutional data, ML can predict stroke (Asadi et al. 2014). A promising method for increasing the precision of forecasting is machine learning, which will help choose the best course of treatment (Tozlu et al. 2020). Clinical decision-making prior to imaging assessment can be aided by an ML model that predicts positive outcomes for acute stroke patients receiving endovascular therapy based on clinical variables (Yoshie et al. 2024).

5. Integration of IoT-ML for Applications in Stroke

Naive Bayes, Decision Trees, Support Vector Machines, Logistic Regression, Linear Regression, Nearest Neighbor, Neural networks, Random Forest, Gradient Boosting,

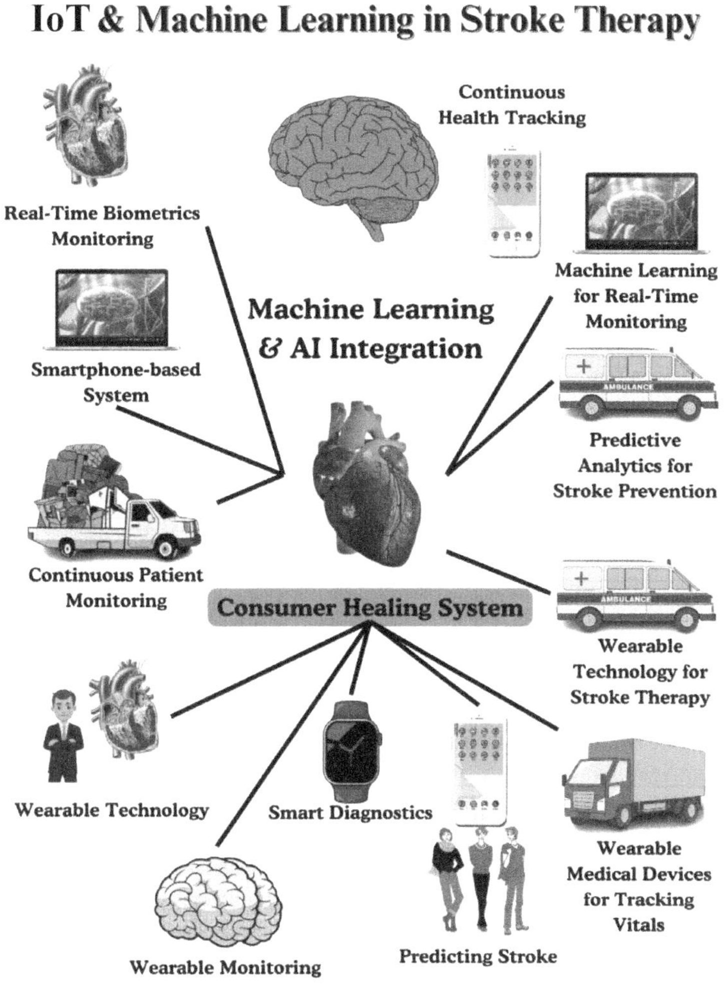

Figure 2: Machine Learning in Stroke Treatment.

Dimensionality Reduction, and K-Means Clustering are utilized in stroke prevention, diagnosis, and treatment (Table 1).

IoT is revolutionizing a number of industries, including healthcare. SmartPants is an IoT wireless device for remote lower limb therapy for stroke victims. It provides a software platform for tracking exercises and nodes for recording physical therapy (Bisio et al. 2019). IoT accurately identifies nine gestures and mimics real-time hand movements for stroke patients (Yang et al. 2018). IoT can predict stroke onset by monitoring physiological parameters and risk factors (Subramaniyam et al. 2018). Numerous research opportunities, such as ML and DL techniques for intelligent stroke rehabilitation, are presented by the Internet of Things' explosive expansion (Razfar et al. 2021). Emotion analysis and management are appealing in IoMT applications and smart wearable sensors because studies have shown that stress and depression can cause stroke (AbdElnapi et al. 2020). Modern-day cities face heat

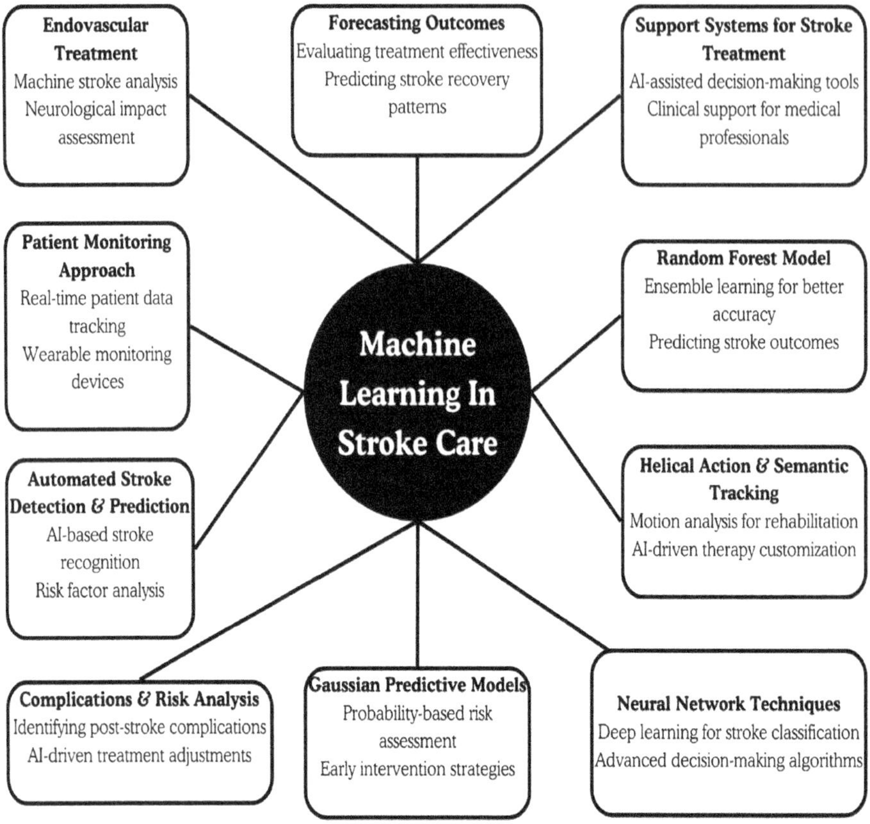

Figure 3: IoT and ML Applications in Stroke.

MOTIMOVE (Tyromotion GmbH) (Jafari et al. 2024), Bioness L300 Go (Kadous 2018) and HandTutor (MediTouch) (Carmeli et al. 2011). With enjoyable, gamified rehabilitation activities, the Neofect Smart Glove (Neofect) assists stroke victims in regaining hand function. The SaeboGlove, a hand rehabilitation assistive glove made by Saebo, Inc., offers passive support to increase grip strength and finger extension. A functional electrical stimulation device for the upper and lower limbs, MOTIMOVE (Tyromotion GmbH) aids in muscle re-education following a stroke. The Bioness L300 Go is a wireless functional electrical stimulation device intended to help stroke patients use their lower limbs more freely and walk less stoically. The HandTutor (MediTouch) is an interactive hand rehabilitation tool that uses biofeedback and motion capture to provide targeted treatment.

Some examples of IoT-based fall detection systems are: Lifeline (Philips Healthcare) (Norlock et al. 2024), Bay Alarm Medical (Ali et al. 2018), Medical Guardian (Islam et al. 2024), GreatCall Lively Mobile Plus (Tedesco et al. 2017), and UnaliWear Kanega Watch. Lifeline (Philips Healthcare) provides medical intervention systems that can sense falls and provide automated notifications.

Table 2: Types of IoT devices used for diagnosis and monitoring of stroke.

Sl. No.	IoT devices	Diagnosis and monitoring of stroke	References
1	Smartwatches	Smartwatches are being used in stroke research for various purposes, including heart rate tracking, AF identification, estimating circadian heart rate rhythm, detecting AF, monitoring life quality and anxiety, controlling aerobic exercise, promoting rehabilitation gaming, monitoring social interactions, and detecting AF. They also offer real-time heart rate monitoring, tennis stroke detection, and remote AF monitoring in cryptogenic strokes. Additionally, smartwatches are being used for psychosocial assessments and device warnings.	(Meza et al. 2023, Ding et al. 2022, Straeten et al. 2023, Lawrie et al. 2018, Dickson et al. 2021, Zhang et al. 2020, Ding et al. 2021, Chandel et al. 2024, Patel and Tarakji 2021, Yao et al. 2024, Taghavi et al. 2019, Ding 2021, Adeboye et al. 2024, Rissardo and Caprara 2021, Naeem et al. 2022, Tanaka et al. 2021, Filippaios et al. 2022, Wouters et al. 2022, Paul et al. 2022, Caprara and Rissardo 2021, Pagola and Meza 2023)
2	Activity trackers	Activity trackers are being used for post stroke monitoring	(Clay et al. 2019, Hargest 2019)
3	Fall detection sensors	Fall detection sensors are being used for prevention of fall of post-stroke patients	(Taylor-Piliae et al. 2016)
4	Motion tracking systems	Motion tracking systems are used for assessment, training, recovery and tracking of stroke patients	(Fan et al. 2020, Nguyen et al. 2024, Bai et al. 2019, Kim et al. 2020)
5	Robotic assistive devices	Robotic assistive devices have shown potential in post-stroke rehabilitation, particularly in hand dexterity, mobility, and upper extremity skills. Early inpatient rehabilitation improved motor recovery, while soft robotic gloves combined with gaming improved hand function. Repetitive task repetition with robotic support improved gait and balance in subacute stroke patients.	(Masiero et al. 2014, Prange-Lasonder et al. 2017, Frick and Alberts 2006, Morone et al. 2016, McConnell et al. 2014, Annanto et al. 2024)
6	Virtual reality systems	Virtual reality systems are used for rehabilitation of stroke patients	(Prashun et al. 2010, Diriba Kenea et al. 2024, Fernandez-Vazquez et al. 2022, Voinea et al. 2015, Micallef 2014)
7	Transcranial magnetic stimulation	Transcranial magnetic stimulation is used for rehabilitation of stroke patients	(Fisicaro et al. 2019, Karatzetzou et al. 2022, Dionisio et al. 2018, Hsu et al. 2012, Kim et al. 2020, He et al. 2020, Zhou et al. 2023, Lefaucheur 2006, Hoyer and Celnik 2011, Xing et al. 2023)

Table 3: Examples of IoT devices used for diagnosis and monitoring of stroke.

Sl. No.	Examples of IoT devices used for diagnosis and monitoring of stroke	Description	References
1	Fitbit (Google)	Wearable fitness trackers are being developed to prevent strokes by monitoring activity, heart rate, and sleep.	(Hui et al. 2018)
2	Apple Watch (Apple Inc.)	For early stroke risk identification, the system provides sophisticated health tracking features including ECG, fall detection, emergency SOS, and abnormal heart rhythm warnings.	(Feldman et al. 2022)
3	Garmin Wearables (Garmin Ltd.)	Activity trackers are devices that monitor various health indicators such as stress, heart rate, and sleep.	(Watts et al. 2024)
4	Biovotion (acquired by Biofourmis)	In order to provide stroke patients with real-time insights into vital indicators, wearable biosensors are being created for ongoing health monitoring.	(Klaas et al. 2021)
5	Oura Ring (Oura Health Ltd.)	A smart ring is being developed that monitors body temperature, heart rate variability, and sleep patterns to detect abnormalities linked to stroke risk.	(Kinnunen et al. 2020)
6	Philips Biosensor BX100 (Philips Healthcare)	Vital indicators including heart rate, skin temperature, and respiration rate are continually monitored via a wearable sensor.	(Santos et al. 2023)
7	VitalConnect (VitalPatch)	The wearable patch is designed for real-time patient monitoring, making it ideal for hospital-at-home situations and post-stroke treatment.	(Leenen et al. 2020)
8	Zio Patch (iRhythm Technologies)	Atrial fibrillation, a major risk factor for stroke, is identified with a continuous heart rhythm monitoring device.	(Tung et al. 2015)
9	Empatica Embrace2 (Empatica Inc.)	A medical-grade wearable is being developed for stroke survivors to monitor physiological stress and seizure activity.	(Rukasha 2022)
10	Preventice BodyGuardian (Boston Scientific)	Real-time ECG data from remote cardiac monitoring equipment is used to identify arrhythmias that might cause a stroke.	(Spooner et al. 2024)
11	ReWalk Robotics	Robotic exoskeletons can enhance gait recovery in stroke patients through rigorous therapy sessions.	(Awad et al. 2020)
12	Ekso Bionics	Exoskeletons provide adaptive support based on the patient's development and are being utilized for gait training in stroke rehabilitation.	(Brenner 2016)
13	Hocoma (DIH Group)	Robotic rehabilitation equipment like Lokomat is used for gait treatment, enabling repeated, task-specific training to improve walking function.	(Soangra et al. 2021)
14	Cyberdyne HAL (Hybrid Assistive Limb)	To help stroke victims regain their motor function, a wearable robotic suit is being created.	(Matsuda et al. 2018)
15	Myomo MyoPro	Powered orthoses aid stroke patients in regaining functional independence by enabling them to move their hands and arms.	(Chang et al. 2024)

Bay Alarm Medical provides wearable alert devices that include GPS tracking for quick fall detection. Fall detection and GPS precision have been improved by the GreatCall Lively Mobile Plus, which is an emergency response tool. GPS tracking, medication reminders, and fall detection are some of the features of the UnaliWear Kanega Watch. Fall detection has been improved for stroke patients by Medical Guardian.

There are many IoT-based healthcare alert systems which include: QardioCore (Qardio, Inc.) (Jaafar and Salam 2019), Cardiogram App (Yazdanian et al. 2016), AngelSense GPS Tracker (Hegde et al. 2019), and MedM Platform (Palmieri 2021). QardioCore (Qardio, Inc.) helps detect arrhythmias. The Cardiogram app identifies atrial fibrillation symptoms. The AngelSense GPS Tracker is useful for stroke patients. MedM Platform combines information from several devices to offer help to stroke patients.

8. Conclusion

Internet of Things devices such as wearable sensors and remote monitoring systems are useful for stroke monitoring and detection. IoT with machine learning in stroke-related applications has proved fruitful. Cloud-based frameworks are essential for this. Machine learning techniques are useful to evaluate risks, create prediction models, and personalize therapies for stroke patients. Supervised, unsupervised learning, deep learning methods, and machine learning techniques are essential for such an approach. There is potential future path for IoT-ML systems in stroke care. Blockchain, machine learning, and artificial intelligence are essential for this approach. IoT and ML work together to improve stroke treatment.

References

AbdElnapi, N. M., Omran, N. F., Ali, A. A., and Omara, F. A. (2020). A framework for stroke prevention using IoT healthcare sensors. in internet of things—applications and future: *Proceedings of ITAF 2019* (pp. 175–186). Springer Singapore.

Aboonq, M. S. (2024). Potential role of logistic regression analysis to identify significant risk factors associated with stroke. *Bulletin of Egyptian Society for Physiological Sciences*, 44(1): 17–28.

Abosede, S. A., Adetunmbi, A. O., and Sarumi, O. A. (2021, December). On-demand data analytics support for hemorrhagic stroke patients using wearable IoT device and fog computing technology. *In International Conference on Soft Computing and Pattern Recognition* (pp. 404–412). Cham: Springer International Publishing.

Adeboye, O., Cooke, G., White, K., Tate, S., Zafonte, R., Narayanan, S. et al. (2024). Abstract TP14: SocialBit: A novel smartwatch sensor to detect social interaction frequency in stroke survivors. *Stroke*, 55(Suppl_1), ATP14-ATP14.

Aghdam, Z. N., Rahmani, A. M., and Hosseinzadeh, M. (2021). The role of the Internet of Things in healthcare: Future trends and challenges. *Computer Methods and Programs in Biomedicine*, 199: 105903.

Ahmadi, H., Arji, G., Shahmoradi, L., Safdari, R., Nilashi, M., and Alizadeh, M. (2019). The application of internet of things in healthcare: a systematic literature review and classification. *Universal Access in the Information Society*, 18: 837–869.

Ain, K., Hidayati, H. B., and Nastiti, O. A. (2020). Expert system for stroke classification using naive bayes classifier and certainty factor as diagnosis supporting device. In *Journal of Physics: Conference Series* (Vol. 1445, No. 1, p. 012026). IOP Publishing.

Fernandez-Lozano, C., Hervella, P., Mato-Abad, V., Rodríguez-Yáñez, M., Suárez-Garaboa, S., López-Dequidt, I. et al. (2021). Random forest-based prediction of stroke outcome. Scientific Reports, 11(1): 10071.

Fernandez-Vazquez, D., Cano-de-la-Cuerda, R., and Navarro-Lopez, V. (2022). Haptic glove systems in combination with semi-immersive virtual reality for upper extremity motor rehabilitation after stroke: A systematic review and meta-analysis. *International Journal of Environmental Research and Public Health*, 19(16): 10378.

Feyen, L., Schott, P., Ochmann, H., Katoh, M., Haage, P., and Freyhardt, P. (2022). Value of machine learning to predict functional outcome of endovascular treatment for acute ischaemic stroke of the posterior circulation. *The Neuroradiology Journal*, 35(3): 363–369.

Filippaios, A., Tran, K. V. T., Mehawej, J., Ding, E., Saczynski, J., Lessard, D. et al. (2022). Abstract P046: Psychosocial Measures In Relation To Smartwatch Device Alerts For Atrial Fibrillation Detection Post Stroke: The Pulsewatch Study. Circulation, 145(Suppl_1), AP046-AP046.

Fisicaro, F., Lanza, G., Grasso, A. A., Pennisi, G., Bella, R., Paulus, W. et al. (2019). Repetitive transcranial magnetic stimulation in stroke rehabilitation: review of the current evidence and pitfalls. *Therapeutic Advances in Neurological Disorders*, 12: 1756286419878317.

Frick, E. M., and Alberts, J. L. (2006). Combined use of repetitive task practice and an assistive robotic device in a patient with subacute stroke. *Physical Therapy*, 86(10): 1378–1386.

GholamAzad, M., Pourmahmoud, J., Atashi, A. R., Farhoudi, M., and Anvari, R. D. (2022). Predicting of stroke risk based on clinical symptoms using the logistic regression method. *International Journal of Industrial Mathematics*, 14(2).

Guhdar, M., Melhum, A. I., and Ibrahim, A. L. (2023). Optimizing accuracy of stroke prediction using logistic regression. *Journal of Technology and Informatics* (JoTI), 4(2): 41–47.

Gupta, R. (2023). Role of machine learning algorithms in stroke diagnosis and treatment. *Proceedings of the Texas AandM Medical Student Grand Rounds*.

Habibzadeh, H., Dinesh, K., Shishvan, O. R., Boggio-Dandry, A., Sharma, G., and Soyata, T. (2019). A survey of healthcare Internet of Things (HIoT): A clinical perspective. *IEEE Internet of Things Journal*, 7(1): 53–71.

Hao, X., Zhang, G. and Ma, S. (2016). Deep learning. *International Journal of Semantic Computing*, 10(03): 417–439.

Hargest, C. E. (2019). A Physiotherapist Facilitated Walking Intervention Using an Activity Tracker to Improve Daily Step Counts in People with Stroke: A Thesis Presented for the Degree of Master of Physiotherapy, University of Otago Dunedin, New Zealand (Doctoral dissertation, University of Otago).

Hawari, H. F., and Yeng, R. B. I. (2023, September). Development of An IoT enabled lower limb stroke rehabilitation system. In *2023 IEEE International Conference on Sensors and Nanotechnology (SENNANO)* (pp. 53–56). IEEE.

He, Y., Li, K., Chen, Q., Yin, J., and Bai, D. (2020). Repetitive transcranial magnetic stimulation on motor recovery for patients with stroke: a PRISMA compliant systematic review and meta-analysis. *American Journal of Physical Medicine and Rehabilitation*, 99(2): 99–108.

Hegde, N., Muralidhara, S., and Ashoka, D. V. (2019). A low-cost and autonomous tracking device for Alzheimer's patients. *Journal of Enabling Technologies*, 13(4): 201–211.

Hoyer, E. H., and Celnik, P. A. (2011). Understanding and enhancing motor recovery after stroke using transcranial magnetic stimulation. *Restorative Neurology and Neuroscience*, 29(6): 395–409.

Hsu, W. Y., Cheng, C. H., Liao, K. K., Lee, I. H., and Lin, Y. Y. (2012). Effects of repetitive transcranial magnetic stimulation on motor functions in patients with stroke: A meta-analysis. *Stroke*, 43(7): 1849–1857.

Hui, J., Heyden, R., Bao, T., Accettone, N., McBay, C., Richardson, J. et al. (2018). Validity of the Fitbit One for measuring activity in community-dwelling stroke survivors. *Physiotherapy Canada*, 70(1): 81–89.

Islam, U., Mehmood, G., Al-Atawi, A. A., Khan, F., Alwageed, H. S., and Cascone, L. (2024). NeuroHealth guardian: A novel hybrid approach for precision brain stroke prediction and healthcare analytics. *Journal of Neuroscience Methods*, 409: 110210.

Iswanto, I., Tulus, T., and Sihombing, P. (2021). Comparison of distance models on K-Nearest Neighbor algorithm in stroke disease detection. *Applied Technology and Computing Science Journal*, 4(1): 63–68.

Jaafar, R., and Salam, A. S. A. (2019, October). Portable electrocardiography with cloud based features: A review of current technologies. In *2019 International Biomedical Instrumentation and Technology Conference (IBITeC)* (Vol. 1, pp. 118–122). IEEE.

Jafari, E., Metani, A., Moreau, B., Nollot, R., Feppon, N., Vergnes-Blanquer, L. et al. (2024). AI-Powered FES-Walking for Gait Rehabilitation After Stroke. In IFFESS 2024 International Conference, 1st to 3rd September 2024, Bath, United Kingdom. *International Functional Electrical Stimulation Society (IFESS)*.

Jang, S. K., Chang, J. Y., Lee, J. S., Lee, E. J., Kim, Y. H., Han, J. H. et al. (2020). Reliability and clinical utility of machine learning to predict stroke prognosis: comparison with logistic regression. *Journal of Stroke*, 22(3): 403.

Javed, S., Ghazala, S., and Faseeha, U. (2020). Perspectives of heat stroke shield: An IoT based solution for the detection and preliminary treatment of heat stroke. *Engineering, Technology and Applied Science Research*, 10(2): 5576–5580.

Jiang, Y., Dang, Y., Wu, Q., Yuan, B., Gao, L., and You, C. (2024). Using a k-means clustering to identify novel phenotypes of acute ischemic stroke and development of its Clinlabomics models. *Frontiers in Neurology*, 15: 1366307.

Jordan, M. I., and Mitchell, T. M. (2015). Machine learning: Trends, perspectives, and prospects. *Science*, 349(6245): 255–60.

Kadous, C. (2018). The Utilization of Bioness L300+ as an Adjunct to Physical Therapy Treatment for a Patient with Left Hemiparesis: A Case Report (Doctoral dissertation, University of Iowa).

Karaca, Y., Cattani, C., Moonis, M., and Bayrak, Ş. (2018). Stroke subtype clustering by multifractal bayesian denoising with fuzzy C Means and K-Means Algorithms. *Complexity*, 2018(1): 9034647.

Karatzetzou, S., Tsiptsios, D., Terzoudi, A., Aggeloussis, N., and Vadikolias, K. (2022). Transcranial magnetic stimulation implementation on stroke prognosis. *Neurological Sciences*, 43(2): 873–888.

Karthick, G. S., and Pankajavalli, P. B. (2020). A review on human healthcare internet of things: a technical perspective. *SN Computer Science*, 1(4): 198.

Khan, M., Jan, B., Farman, H., Ahmad, J., Farman, H. and Jan, Z. (2019). Deep learning methods and applications. In: *Deep Learning: Convergence to Big Data Analytics*, 31–42. Spinger Link.

Khorsand, B., Vaghf, A., Salimi, V., Zand, M., and Ghoreishi, S. A. (2024). Enhancing Ischemic Stroke Management: Leveraging Machine Learning Models for Predicting Patient Recovery After Alteplase Treatment. medRxiv, 2024-11.

Kim, H. (2022). Deep learning. *In*: Artificial Intelligence for 6G (pp. 247–303). Cham: Springer International Publishing.

Kim, H., Lee, A., Shin, Y. I., Chang, W. H., Koo, K. H., Seong, H., and Kim, Y. H. (2018). Effects of digital smart glove system on motor recovery of upper extremity in subacute stroke patients. *Annals of Physical and Rehabilitation Medicine*, 61: e28.

Kim, S., Park, S., and Lee, O. (2020). Development of a diagnosis and evaluation system for hemiplegic patients post-stroke based on motion recognition tracking and analysis of wrist joint kinematics. *Sensors*, 20(16): 4548.

Kim, W. J., Rosselin, C., Amatya, B., Hafezi, P., and Khan, F. (2020). Repetitive transcranial magnetic stimulation for management of post-stroke impairments: An overview of systematic reviews. *Journal of Rehabilitation Medicine*, 52(2): 1–10.

Kinnunen, H., Rantanen, A., Kenttä, T., and Koskimäki, H. (2020). Feasible assessment of recovery and cardiovascular health: accuracy of nocturnal HR and HRV assessed via ring PPG in comparison to medical grade ECG. Physiological Measurement, 41(4): 04NT01.

Klaas, V. C., Calatroni, A., Pavic, M., Guckenberger, M., Theile, G., and Tröster, G. (2021). Monitoring Patients in Ambulatory Palliative Care: a Design for an Observational Study. *EAI Endorsed Transactions on Mobile Communications and Applications*, 6(17): e3–e3.

Kulkarni, A., and Sathe, S. (2014). Healthcare applications of the Internet of Things: A Review. *International Journal of Computer Science and Information Technologies*, 5(5): 6229–6232.

Lalitha, K., Kumar, D. R., Poongodi, C., and Arumugam, J. (2021). Healthcare internet of things–the role of communication tools and technologies. In *Blockchain, Internet of Things, and Artificial Intelligence* (pp. 331–348). Chapman and Hall/CRC.

Laplante, P. A., and Laplante, N. (2016). The internet of things in healthcare: Potential applications and challenges. *It Professional*, 18(3): 2–4.

Lauzon, F.Q. (Jul. 2012). An introduction to deep learning. *In*: 2012 11th International

Lawrie, S., Dong, Y., Steins, D., Xia, Z., Esser, P., Sun, S. et al. (2018). Evaluation of a smartwatch-based intervention providing feedback of daily activity within a research-naive stroke ward: a pilot randomised controlled trial. *Pilot and Feasibility Studies*, 4: 1–18.

LeCun, Y., Bengio, Y. and Hinton, G. (2015). Deep learning. *Nature*, 521(7553): 436–44.

Leenen, J. P., Leerentveld, C., van Dijk, J. D., van Westreenen, H. L., Schoonhoven, L., and Patijn, G. A. (2020). Current evidence for continuous vital signs monitoring by wearable wireless devices in hospitalized adults: systematic review. *Journal of Medical Internet Research*, 22(6): e18636.

Lefaucheur, J. P. (2006). Stroke recovery can be enhanced by using repetitive transcranial magnetic stimulation (rTMS). *Neurophysiologie Clinique/Clinical Neurophysiology*, 36(3): 105–115.

Liu, Y., Ma, B., and Wang, Y. (2021, December). Study on prediction model of stroke risk based on decision tree and regression model. In *2021 IEEE International Conference on Big Data (Big Data)* (pp. 4798–4801). IEEE.

Ma, Z., Yue, Y., Luo, Y., Wang, W., Cao, Y., and Fang, Q. (2020). Clinical utility of the inflammatory factors combined with lipid markers in the diagnostic and prognostic assessment of ischemic stroke: based on logistic regression models. *Journal of Stroke and Cerebrovascular Diseases*, 29(4): 104653.

MacDermott, Á., Kendrick, P., Idowu, I., Ashall, M., and Shi, Q. (2019, June). Securing things in the healthcare internet of things. In *2019 Global IoT Summit (GIoTS)* (pp. 1–6). IEEE.

Mahesh, B. (2020). Machine learning algorithms: A review. *International Journal of Science and Research (IJSR)*. [Internet], 9(1): 381–86.

Mahmoud, H., Aljaldi, F., El-Fiky, A., Battecha, K., Thabet, A., Alayat, M. et al. (2023). Artificial Intelligence machine learning and conventional physical therapy for upper limb outcome in patients with stroke: a systematic review and meta-analysis. *European Review for Medical and Pharmacological Sciences*, 27(11).

Maksimović, M., Vujović, V., and Perišić, B. (2015, June). A custom Internet of Things healthcare system. *In 2015 10th iberian conference on information systems and technologies (CISTI)* (pp. 1–6). IEEE.

Masiero, S., Armani, M., Ferlini, G., Rosati, G., and Rossi, A. (2014). Randomized trial of a robotic assistive device for the upper extremity during early inpatient stroke rehabilitation. *Neurorehabilitation and Neural Repair*, 28(4): 377–386.

Mathew, A., Amudha, P., and Sivakumari, S. (2021). Deep learning techniques: An overview. Advanced *Machine Learning Technologies and Applications: Proceedings of AMLTA* 2020, 599–608.

Matsuda, M., Iwasaki, N., Mataki, Y., Mutsuzaki, H., Yoshikawa, K., Takahashi, K. et al. (2018). Robot-assisted training using Hybrid Assistive Limb® for cerebral palsy. *Brain and Development*, 40(8): 642–648.

Mazlan, S., Rahman, H. A., Kader, B. S. K., Yeong, C. F., and Alhusni, N. A. M. R. (2021). Multiple linear regression in predicting motor assessment scale of stroke patients. *International Journal of Integrated Engineering*, 13(6): 330–338.

McConnell, A., Kong, X., and Vargas, P. A. (2014, August). A novel robotic assistive device for stroke-rehabilitation. In *The 23rd IEEE International Symposium on Robot and Human Interactive Communication* (pp. 917–923). IEEE.

Meza, C., Juega, J., Francisco, J., Santos, A., Duran, L., Rodriguez, M. et al. (2023). Accuracy of a smartwatch to assess heart rate monitoring and atrial fibrillation in stroke patients. *Sensors*, 23(10): 4632.

Miao, S., Shen, C., Feng, X., Zhu, Q., Shorfuzzaman, M., and Lv, Z. (2021). Upper limb rehabilitation system for stroke survivors based on multi-modal sensors and machine learning. *IEEE Access*, 9: 30283–30291.

Micallef, L. (2014). Effectiveness of virtual reality systems in the rehabilitation process of adult stroke patients (Bachelor's thesis, University of Malta).

Mitchell, T. M. (2006). The Discipline of Machine Learning. (Vol. 9, p. 3). Pittsburgh: Carnegie Mellon University, School of Computer Science, Machine Learning Department. *Research Gate.*

Mitra, J., Bourgeat, P., Fripp, J., Ghose, S., Rose, S., Salvado, O. et al. (2014). Lesion segmentation from multimodal MRI using random forest following ischemic stroke. *NeuroImage*, 98: 324–335.

Morone, G., Annicchiarico, R., Iosa, M., Federici, A., Paolucci, S., Cortés, U., and Caltagirone, C. (2016). Overground walking training with the i-Walker, a robotic servo-assistive device, enhances balance in patients with subacute stroke: a randomized controlled trial. *Journal of Neuroengineering and Rehabilitation*, 13: 1–10.

Naeem, S., Tran, K. V., Ding, E., Mehawej, J., Wang, Z., Lessard, D. et al. (2022). assessing quality of life outcomes among elderly population of smartwatch users monitored for post-stroke atrial fibrillation: the pulsewatch study. *Journal of the American College of Cardiology*, 79(9_Supplement): 2051–2051.

Nalini, D., and Periasamy, R. (2015). Lloyd and minkowski based K-Means clustering for effective diagnosis of heart disease and stroke. *International Review*, 10(6): 573–579.

Naresh, V. S., Pericherla, S. S., Murty, P. S. R., and Reddi, S. (2020). Internet of Things in healthcare: Architecture, applications, challenges, and solutions. *Computer Systems Science and Engineering*, 35(6).

Nguyen, T. T., Nguyen, Q. D., Ha, M. Q., Hoang, X. T., Bui, T. N., and Tien, D. L. (2024). The development of a motion-tracking system to assess the recovery level for stroke survivors. *Acta Polytechnica Hungarica*, 21(9): 9–28.

Nie, X., Yang, J., Li, X., Zhan, T., Liu, D., Yan, H. et al. (2024). Prediction of futile recanalisation after endovascular treatment in acute ischaemic stroke: development and validation of a hybrid machine learning model. *Stroke and Vascular Neurology*, svn-2023.

Ningrum, R. F., Yosrita, E., Djunaidi, K., Siregar, R. R. A., and Anitasari, Y. (2023, August). Implementation Data Mining with the Naive Bayes Classifier Algorithm in Determining the Type of Stroke. In 2023 *International Conference on Networking, Electrical Engineering, Computer Science, and Technology (IConNECT)* (pp. 247–252). IEEE.

Norlock, V., Vazquez, R., Dunn, A., Siegfried, C., Wadhwa, M., and Medic, G. (2024). Comparing the outcomes and costs of cardiac monitoring with implantable loop recorders and mobile cardiac outpatient telemetry following stroke using real-world evidence. *Journal of Comparative Effectiveness Research*, 13(6): e240008.

Ogasawara, T., Mukaino, M., Matsunaga, K., Wada, Y., Suzuki, T., Aoshima, Y. et al. (2024). Prediction of stroke patients' bedroom-stay duration: machine-learning approach using wearable sensor data. *Frontiers in Bioengineering and Biotechnology*, 11: 1285945.

Pagola, J., and Meza, C. M. B. (2023). Accuracy of real time heart rate monitoring using smartwatch in a stroke care unit. *Europace, 25(Supplement_1), euad122-572.*

Palmieri, F. (2021). Multi-sided non transactional platforms in healthcare: Value Capture from Real-World Data.

Park, J. D., Kim, Y. Y., and Lee, C. Y. (2009). Identification of epistasis in ischemic stroke using multifactor dimensionality reduction and entropy decomposition. *BMB Reports*, 42(9): 617–622.

Park, S. J., Hussain, I., Hong, S., Kim, D., Park, H., and Benjamin, H. C. M. (2020, January). Real-time gait monitoring system for consumer stroke prediction service. *In 2020 IEEE International Conference on Consumer Electronics (ICCE)* (pp. 1–4). IEEE.

Patel, D., and Tarakji, K. G. (2021). Smartwatch diagnosis of atrial fibrillation in patient with embolic stroke of unknown source: a case report. *Cardiovascular Digital Health Journal*, 2(1): 84–87.

Paul, T. J., Tran, K. V. T., Mehawej, J., Lessard, D., Filippaios, A., Ding, E. et al. (2022). Abstract P045: Self-reported Anxiety and Patient Activation Among Smartwatch Users Monitored For Atrial Fibrillation Post-stroke: The Pulsewatch Study. *Circulation, 145(Suppl_1)*: AP045-AP045.

Phuc, P. H., Toan, V. D., Ha, T. Q., Nghia, V. H., Son, B. C., and Phuc, T. D. (2023, December). High accurate stroke prediction smart necklace device based on IoT and machine learning algorithm. In *2023 1st International Conference on Health Science and Technology (ICHST)* (pp. 1–6). IEEE.

Postolache, O., Hemanth, D. J., Alexandre, R., Gupta, D., Geman, O., and Khanna, A. (2020). Remote monitoring of physical rehabilitation of stroke patients using IoT and virtual reality. *IEEE Journal on Selected Areas in Communications*, 39(2): 562–573.

Tung, C. E., Su, D., Turakhia, M. P., and Lansberg, M. G. (2015). Diagnostic yield of extended cardiac patch monitoring in patients with stroke or TIA. *Frontiers in Neurology*, 5: 266.

Vajargah, K. F., Sadeghi-Bazargani, H., Mehdizadeh-Esfanjani, R., Savadi-Oskouei, D., and Farhoudi, M. (2012). OPLS statistical model versus linear regression to assess sonographic predictors of stroke prognosis. *Neuropsychiatric Disease and Treatment*, 387–392.

Van Os, H. J., Ramos, L. A., Hilbert, A., Van Leeuwen, M., Van Walderveen, M. A., Kruyt, N. D. et al. (2018). Predicting outcome of endovascular treatment for acute ischemic stroke: potential value of machine learning algorithms. *Frontiers in Neurology*, 9: 784.

Ventura, R. B., Surano, F. V., and Porfiri, M. (2021, March). Automated classification of bimanual movements in stroke telerehabilitation: A comparison of dimensionality reduction algorithms. *In Nano-, Bio-, Info-Tech Sensors and Wearable Systems* (Vol. 11590, pp. 8–14). SPIE.

Vibin, V. S., Archidha, R., MK, S. S., and Kathiravan, S. (2024, March). Development Of Iot Based Assistive System For Stroke Survivors. *In 2024 Tenth International Conference on Bio Signals, Images, and Instrumentation (ICBSII)* (pp. 1–6). IEEE.

Voinea, A., Moldoveanu, A., and Moldoveanu, F. (2015, May). 3D visualization in IT systems used for post stroke recovery: Rehabilitation based on virtual reality. *In 2015 20th international conference on control systems and computer Science* (pp. 856–862). IEEE.

Wang, H., Ma, C., and Zhou, L. (2009, December). A brief review of machine learning and its application. In: *2009 International Conference on Information Engineering and Computer Science* (pp. 1–4). IEEE.

Watts, S. P., Binnie, M. J., Goods, P. S., Doyle, M. M., Hewlett, J., and Peeling, P. (2024). Garmin wearable device offers reliable alternative for on-water stroke rate and velocity measurement in rowing. Proceedings of the Institution of Mechanical Engineers, Part P: *Journal of Sports Engineering and Technology*, 238(3): 242–250.

Webb, S. (2018). Deep learning for biology. *Nature*, 554(7693): 555–57.

Wouters, F., Gruwez, H., Vranken, J., Vanhaen, D., Daelman, B., Ernon, L. et al. (2022). Aim: This paper presents the preliminary results from the ongoing REMOTE trial. It aims to explore the opportunities and hurdles of using insertable cardiac monitors (ICMs) and photoplethysmography-based mobile health (PPG-based mHealth) using a smartphone or smartwatch to detect atrial fibrillation (AF) in cryptogenic stroke and transient ischemic attack (TIA) patients. *Wearable Devices for Cardiac Rhythm Monitoring*.

Wulandari, E., and Witanti, A. (2024). The classification of stroke prediction using the support vector machine (SVM) method. Jatisi (Jurnal Teknik Informatika dan Sistem Informasi), 11(3).

Xie, Y., Jiang, B., Gong, E., Li, Y., Zhu, G., Michel, P. et al. (2019). Use of gradient boosting machine learning to predict patient outcome in acute ischemic stroke on the basis of imaging, demographic, and clinical information. *American Journal of Roentgenology*, 212(1): 44–51.

Xing, Y., Zhang, Y., Li, C., Luo, L., Hua, Y., Hu, J. et al. (2023). Repetitive transcranial magnetic stimulation of the brain after ischemic stroke: mechanisms from animal models. *Cellular and Molecular Neurobiology*, 43(4): 1487–1497.

Xu, Y., Holanda, G., Souza, L. F. D. F., Silva, H., Gomes, A., Silva, I. et al. (2020). Deep learning-enhanced internet of medical things to analyze brain ct scans of hemorrhagic stroke patients: a new approach. *IEEE Sensors Journal*, 21(22): 24941–24951.

Yang, G., Deng, J., Pang, G., Zhang, H., Li, J., Deng, B. et al. (2018). An IoT-enabled stroke rehabilitation system based on smart wearable armband and machine learning. *IEEE Journal of Translational Engineering in Health and Medicine*, 6: 1–10.

Yang, K. C., Huang, C. H., and ChiangLin, C. Y. (2018, December). Combining IOT and Android APP System for Upper Limb Stroke Rehabilitation. In *2018 IEEE International Conference on Industrial Engineering and Engineering Management (IEEM)* (pp. 1096–1100). IEEE.

Yao, L., Pan, X., Yang, Y., Zhong, W., Li, J., Wang, Z. et al. (2024). Accuracy and reliability of wristband smartwatch-based estimation of circadian rhythm of heart rate in stroke patients. *Neurology Asia*, 29(1).

Yazdanian, H., Mahnam, A., Edrisi, M., and Esfahani, M. A. (2016). Design and implementation of a portable impedance cardiography system for noninvasive stroke volume monitoring. *Journal of Medical Signals and Sensors*, 6(1): 47–56.

Yeole, A. S., and Kalbande, D. R. (2016, March). Use of Internet of Things (IoT) in healthcare: A survey. In *Proceedings of the ACM Symposium on Women in Research 2016* (pp. 71–76).

Yoshie, T., Yoshimura, S., Wada, S., Sakuraba, M., Suzuki, T., Yoko, S. et al. (2024). Abstract WP203: Prediction of Outcome After Endovascular Therapy for Acute Ischemic Stroke: A Machine Learning Approach From Japan Stroke Data Bank. *Stroke*, 55(Suppl_1), AWP203-AWP203.

Yuehong, Y. I. N., Zeng, Y., Chen, X., and Fan, Y. (2016). The internet of things in healthcare: An Overview. *Journal of Industrial Information Integration*, 1: 3–13.

Zhang, J., Bi, H., Chen, Y., Zhang, Q., Fu, Z., Li, Y. et al. (2020). SmartSO: Chinese character and stroke order recognition with smartwatch. *IEEE Transactions on Mobile Computing*, 20(7): 2490–2504.

Zhang, S., Wang, H., Jiang, Y., and Wang, X. (2023, November). Clinical Intelligent Diagnosis and Treatment of Hemorrhagic Stroke Based on Random Forest and BP-Neural Networks. In *2023 3rd International Signal Processing, Communications and Engineering Management Conference (ISPCEM)* (pp. 455–460). IEEE.

Zhang, Y. (2025). Multi-sensors edge computing for tennis stroke recognition in IoT environment. *Internet Technology Letters*, e582.

Zhou, L., Jin, Y., Wu, D., Cun, Y., Zhang, C., Peng, Y. et al. (2023). Current evidence, clinical applications, and future directions of transcranial magnetic stimulation as a treatment for ischemic stroke. *Frontiers in Neuroscience*, 17: 1177283.

CHAPTER 10

Wearable and Implantable Sensors for Real-Time Health Tracking

Shruti Mehra, Avinash Kumar Sharma, Shajee Mohan B. S.* and *Neha Agarwal*

1. Introduction

"IoT" stands for Internet of Things, a structure of interconnected computational devices, machine-driven and automated systems, matters, or individuals that are provided with unique identifiers and the capability to transfer information over a system without requiring humanoid-to-humanoid or humanoid-to-supercomputer interaction. Lifespan expectation has been growing worldwide due to noteworthy enhancements in health care and medication and with rising awareness of individual and environmental hygiene or healthcare with IOMT (Chakraborty 2024). In the last few years, there has been a rapid increase in personal planning, which has led to a reduction in birth rates across the world. According to the WHO, by 2025, the aging population over 65 years old is predicted to outnumber children below 5 years. However, this large aging population is expected to create a significant influence on the financial conditions of society in terms of communal well-being and healthcare needs. This leads to the costs related to healthcare amenities continuing to rise because of the increasing price of medical treatment, equipment, and medicinal care units (Sharma 2024). Therefore, it is essential to develop and implement novel approaches and skills in order to deliver improved medical health services at an affordable cost to the aged population or to individuals in regions with limited access to healthcare while ensuring maximum ease, individuality, and participation among individuals. Remote healthcare tracking allows individuals to continue to stay at home rather than

Department of Computer science and Engineering, Sharda School of Computing Sciences and Engineering (SSCSE), Sharda University, Greater Noida, India.
Emails: avinashsharma2006@gmail.com, shajee.mohan@sharda.ac.in, neha.agarwal1@sharda.ac.in
* Corresponding author: shrutimehra712@gmail.com

DOI: 10.1201/9781003783855-13

in expensive healthcare amenities such as infirmaries or nursing homes. It delivers a well-organized and cost-effective substitute for on-the-spot medical tracking. Such systems, equipped with non-invasive and inconspicuous wearable sensors, will be feasible diagnostic tools for healthcare employees for one-to-one care, monitoring significant physical signs and actions of patients in real time from remote locations. Therefore, it is clear that wearable sensor play a critical role in such tracking systems that have attracted the attention of numerous scholars, businesspersons, and technological companies in recent years. A variety of application-specific wearable sensors, physical and activity tracking systems. Wearable devices can track and store real-time data related to biological conditions and motion activities. Wearable sensor-based well-being tracking systems might include different kinds of flexible sensors that can be integrated into fabric fibers, attires, flexible patches, or directly attached to the human body. The sensors are capable of measuring physical signs such as ECG, EMG, HR, body temperature, EDA, arterial oxygen saturation, SpO2, BP, and respiration rate. In addition, MEMS actuators and sensors such as accelerometers, gyroscopes, and magnetic field sensors are extensively used for measuring activity-related signals. Continuous tracking of physiological signals might help detect and identify numerous vascular, neurological, and pulmonary illnesses at early onset. Also, real-time monitoring of an individual's motion activities might be valuable in detecting gait patterns and posture scrutiny or in drowse valuation. The vesture health tracking systems are typically equipped with a variety of electric and MEMS sensors, actuators, radiocommunication modules, and signal processing units. The data obtained by the sensors linked in a radiocommunication body sensor network (BSN) are conveyed to a nearby processing node using an appropriate communication procedure; first, it prefers a low-power and short-range radiocommunication medium, e.g., Bluetooth, ZigBee, or Near field communication (NFC). The processing unit, which might be a Personal Digital Assistant (PDA), smartphone, computer, or a commissioned processing component based on a microcontroller or a field-programmable gate array (FPGA), performs advanced processing, scrutiny, and decision procedures and may also store and display the results to the user. It conveys patient information on the network to healthcare workers, thus functioning as the gateway to remote healthcare amenities. To be used for long-term monitoring, wearable health tracking systems need to satisfy certain medical and ergonomic requirements. For example, the system needs to be comfortable, the mechanisms must be flexible, small in size, and chemically inert, harmless, and hypoallergenic to humans. Wearable sensors, becoming increasingly more comfortable and less noticeable, are suitable for monitoring a person's well-being without interfering with their daily activities. The sensors can calculate various functional signals/parameters as well as the motions of an individual by employing them at different sites of the body. The progress in small-power, compact wearable, low-cost computing and storage devices combined with contemporary communication technologies paves the way for low-cost, unobtrusive, and long-term health monitoring systems. The use of the IOT in healthcare has rapidly increased across numerous Internet of Things use cases. The growth of healthcare with remote tracking and telemonitoring as main applications. One area where numerous enterprises operate is tracking, monitoring, and maintenance of assets using IOT (Gowda 2024). This applies to both medical interaction with extended battery-operated operation. Two diverse device kinds can contribute in an LR-WPAN system—a full-function device (FFD) and a reduced-function device (RFD). The FFD can function in three modes serving as a PAN manager. An FFD can communicate with an RFD or another FFD, while an RFD can converse only with an FFD. An RFD is designed for very simple requests, such as a light switch or an infrared sensor. They do not need to guide large quantities of information and may associate only with a single FFD at a time. The RFD can be applied using minimal resources and memory volume. After an FFD is triggered for the primary time, it may form its own network and become the PAN manager.

(d) Wi-Fi:

Radiocommunication faithfulness comprises values for Wireless Local Area Networks (WLAN). It permits operators to work on internet at broadband speeds when linked to an access point (AP) or in ad hoc mode. This structure contains numerous mechanisms that interact to deliver a wireless LAN that supports position mobility transparently to higher layers. The basic cell of an IEEE 802.11 LAN is called a Basic Service Set (BSS), which is a set of movable or fixed stations. If a station moves out of its BSS, it can no longer directly connect with other members of the BSS. Based on the BSS services, the independent Basic Service Set (IBSS) and extended service set (ESS) network configurations exist. LAN is often created without pre-planning for as long as an ad hoc network. This type of operation is frequently referred to as a substructure system.

***(e) En-Ocean*:**

En-Ocean is a scheme for transmitting information wirelessly that needs no external power source or maintenance and instead uses energy harvesting technology to produce the minimal amount of energy required from the environment. The frequency differs depending on region. In Japan, this is classified as having low influence on radiocommunication by means of the 315 MHz and 928 MHz incidences. In the US, 315 MHz and 902 MHz are employed, while the EU adopts 868 MHz.

Learning of numerous existing IoT-based well-being schemes presents their compensations. First, we will discuss numerous IoT-based healthcare applications, including both solitary and clustered-procedure requests as follows:

Glucose Level Sensing—Diabetes is a collection of metabolic disorders in which there are high blood glucose levels over an extended period. Blood glucose tracking reveals persons' patterns of blood glucose deviations and assists in planning food intake and actions. An m-IOT formation methodology for noninvasive glucose detection on a real-time basis is proposed. In this technique, devices from patients are linked through IPV6 connectivity to pertinent healthcare practitioners. The efficiency model reveals a broadcast device for the transmission of collected bodily information on plasma glucose based on IoT systems. This includes a blood glucose collector, a mobile phone or a computer, and a background processor. A comparable innovation is emerging. In addition, a general IoT-based medical acquisition sensor can be used to display glucose level.

Electrogram Monitoring—The ECG, that is, the electric activity of the heart logged by electrocardiography, comprises the measurement of simple heart rate and the determination of the basic rhythm as well as the analysis of complex arrhythmias, myocardial ischemia, and prolonged QT intervals. The use of IoT for ECG tracking, which forms IoT-enabled Wireless Body Area Networks (WBAN) (Olatinwo 2025), has the potential to offer extensive data and can be used to its fullest extent. A number of studies have clearly discussed IoT-based ECG tracking. The innovation presents an IoT-based ECG tracking scheme composed of a portable wireless acquisition transmitter and a wireless receiving processor. The scheme integrates a detection automation technique to sense irregular information such that cardiac functions can be recognized on a real-time basis. It follows a comprehensive recognition procedure of ECG signals at the application layer of the IoT system for ECG tracking.

Blood Pressure Monitoring—In this, there is a combination of a KIT blood pressure (BP) meter and an NFC-enabled KIT mobile receiver that becomes part of BP tracking based on IoT and is discussed as the main addressed system. A challenging situation is that BP must be frequently monitored. This is completed using Withing's BP device; this device functions by connecting to a mobile computing system. A device for BP information gathering and transmission over an IoT system is proposed. This device is composed of a BP meter built with a communication module.

Body Temperature Monitoring—The temperature of the body can be tracked and is a vital portion of well-being maintenance facilities since it plays a vital role as a significant vital sign in the maintenance of homeostasis. In the m-IOT perception, it is shown that a sensor embedded in the Telos mote and a characteristic example of measured body temperature variations demonstrates the effective process of the established m-IOT scheme. A temperature measurement device based on a gateway over the IoT is proposed.

Oxygen Saturation Monitoring—Pulse oximetry is appropriate for the non-invasive continuous tracking of blood oxygen capacity. The integration of IoT with pulse oximetry is valuable for technology-driven medical healthcare requests. A review of CoAP-based healthcare facilities discusses the potential of IoT pulse oximetry. The device originates with connectivity built on Bluetooth health device profiles, and the device attaches directly to the Monera platform. An IoT-optimized low-power pulse oximeter for distant patient tracking is proposed. This device can be used to continuously monitor patient health over an IoT network.

Rehabilitation System—This rehabilitation system can improve and reinstate the functional skill and quality of life of persons with bodily impairments or disabilities; they represent a dynamic division of medication. The IoT has the potential to improve rehabilitation schemes in terms of addressing issues related to aging populations. An ontology-based systematic enterprise technique for IoT-based intelligent rehabilitation systems is proposed. This enterprise successfully proves that the IoT can be an effective platform for linking all essential resources to support real-time interaction. IoT-based technologies can provide a valuable structure to support actual distant sessions in wide-ranging rehabilitation.

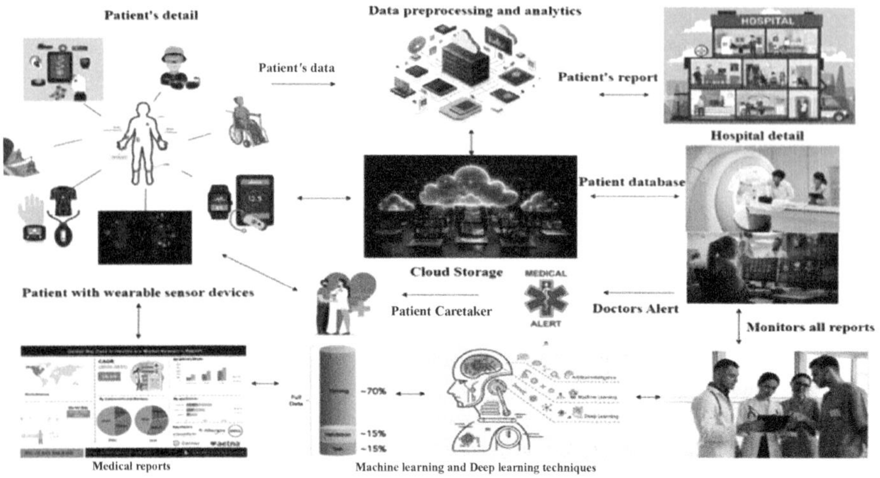

Figure 1: Revolutionizing healthcare with IOMT from patient wearables (Sripathi Raja 2016). This figure depicts that data is collected by sensors and various medical reports, where it can be processed using data preprocessing techniques, analysis, and data visualizations that help clinicians for diagnostic purposes, and this data can be stored over the cloud, which will be beneficial for patients and doctors to view the reports online.

Wheelchair Management—A healthcare scheme for wheelchair users based on IoT technology is proposed. The design originates with WBANs combined with numerous devices whose functions are aligned with IoT provisions. This scheme offers chair vibration control and can detect the status of the wheelchair user. This progress ultimately shows that ordinary "things" can evolve into linked apparatuses driven by data. This device can display essentials of specific sitting positions in the chair and gather data on the user's surroundings, permitting evaluation of a location's accessibility.

2. Literature Survey

The above literature survey depicts various techniques and wearable and implantable devices that are used, and various other feature optimization techniques are used, which can convert raw data into a useful form while avoiding unnecessary information from it, preserving the most relevant features that can be used for the decision-making process, which can be beneficial for clinicians as well as patients for early diagnosis.

3. Demonstration of Wearable and Implantable Devices

The above diagram depicts the block diagram of connecting various sensors like a heart sensor or OpenBCI-compatible heart sensor, pulse sensor, and temperature sensor where they collect the raw data and feed it to the Arduino Uno, which can be displayed on a 16*2 LCD screen with a buzzer sound that indicates an emergency when there is a high pulse or heartbeat rate. Further, we can add more data analysis techniques into it for visualization purposes, such as storing raw data received from

Table 1: Shows the review of various papers with the techniques used alongwith their advantages and limitations.

S. No	Authors & Year	Techniques Used	Advantages	Limitations
1	(Ahmed, 2024)	IoMT, wearable ECG monitors	Enhances cardiac care; continuous monitoring	Network latency; limited battery life
2	(Amit, 2024)	Wearable sensors, mobile app	Ensures elderly medication adherence	Requires smartphone usage; privacy concerns
3	(Damilola, 2025)	IoT-enabled assistive systems	Aids elderly/disabled; smart alerts	High setup complexity and cost
4	(Dankan, 2024)	Automated medicine dispenser, wearables	Reduces medication errors	Limited personalization; risk of failure
5	(Kallur, 2024)	Cyber-Physical-Social Systems	Enhances engagement; holistic system	Data integration and synchronization issues
6	(Manivannam, 2024)	Review on wearables	Summarizes existing tech trends	No practical validation
7	(Manoj, 2025)	Edge AI, wearable sensors	Real-time low-latency health monitoring	Hardware dependency; energy usage
8	(Mohammad, 2023)	Smart wearable sensor review	Future trends and issues discussed	Conceptual only; lacks implementation
9	(Narmatha, 2024)	IoT, Blockchain, IPFS	Secure, immutable data sharing	High computational cost
10	(Nithyavani, 2016)	Review of smart IoT systems	Broad IoT applications covered	Lacks implementation depth
11	(Prachi, 2024)	ML-integrated IoT diagnostics	Personalized monitoring and prediction	Needs big data; risk of bias in ML
12	(Rahul, 2025)	Interpretable ML, WBAN	Transparent heart disease prediction	Signal interference in WBAN; power usage
13	(Rishabh, 2024)	Deep Learning and Dimensionality Reduction	Early diabetes detection; faster models	High computation; sensitive to input data
14	(Rujul, 2025)	CIoT and DRNN-LSTM	Secure framework with predictive DL	Security overhead adds latency
15	(Sandeep, 2020)	Real-time cardiac IoT system	Simple and real-time cardiac alerts	Outdated model; lacks adaptive AI
16	(Sathiya, 2025)	IoT and Modified Neural Network	Good accuracy in heart prediction	Complex model; integration issues
17	(Saurabh, 2024)	MSSO-ANFIS, IoMT	Adaptive, intelligent heart monitoring	Complex fuzzy rule tuning
18	(Sripathi, 2024)	GNN + Leopard Seal Optimization	Accurate automated diagnosis	Limited real-world testing
19	(Sudhanshu, 2024)	Feature Optimization and Deep Learning	Precise early heart disease prediction	Data dependency; high computation
20	(Thavavel, 2024)	Machine Learning and Monarch Butterfly Optimization	Novel optimizer improves results	Needs validation vs standard methods
21	(Wahidur, 2023)	Optimized CNN for ECG signals	High ECG classification accuracy	Real-time tuning needed

research and advanced medical applications like robotic limb control in paralyzed patients.

- **Semi-Invasive BCI Sensors** – Positioned inside the skull but outside brain tissue, such as electrocorticography (ECoG) grids, offering a balance between signal quality and safety.
- **Non-Invasive BCI Sensors** – These types of sensors are on the scalp without surgery, the most of them are EEG electrodes. It uses MSSO-ANFIS for accurate and efficient decision-making by clinicians. (Khan 2023). These are broadly used for real-time tracking of brain signals, cognitive workload assessment, and neurofeedback therapies.

*3. **EEG-Based BCI Sensors***

EEG-based BCI sensors are the most extensively used for real-time health monitoring due to their portability, safety, and cost-effectiveness. These sensors detect voltage fluctuations caused by ionic currents in brain neurons. EEG BCIs can operate in various patterns:

- **Event-Related Potentials (ERP)** – Responses to specific stimuli, such as the P300 wave, useful for communication interfaces.
- **Motor Imagery (MI)** – Neural activity patterns generated when a user imagines movement, enabling control of assistive devices.
- **Steady-State Visually Evoked Potentials (SSVEP)** – Brain responds to various visual stimuli, as brain responds to visual effects more than audio at specific frequencies, offering high accuracy in device control.

*4. **Applications in Real-Time Health Tracking***

BCI sensors have broad applications in neurology, rehabilitation, and assistive technology:

- **Neuro prosthetics** – It can be used in real time for controlling robotic arms, wheelchairs, or any other devices directly from brain signals.
- **Communication Aids** – Empowering persons with locked-in syndrome to communicate using thought-based interfaces that reflect what thoughts are created in the mind.
- **Seizure Prediction** – These sensors are used for predicting seizures and alerting about these seizure attacks.
- **Cognitive and Emotional Monitoring** – These types of sensors are used to track the emotional level or pressure caused by emotional stress, or occupational workload, which can be fatal if not tracked early, as it may lead to suicide cases. So, early monitoring results in proper diagnosis.

3.2 Arduino Uno: Overview of the Arduino Uno

The ATmega328P serves as the basis for the Arduino Uno microcontroller board. In the present scenario, it is one of the most important and user-friendly expansion boards in the Arduino family, widely used for automation, Internet of Things applications, and electronics projects. The board consists of six analog inputs, fourteen digital input/output pins, a power jack, an ICSP header, a USB port, a reset button, and a quartz crystal operating at 16 MHz.

It runs on 5V and can be powered with an external supply (7–12V is advised) or a USB cable. The Arduino IDE, which supports C and C++, is used to program the Arduino Uno. It is suitable for both novices and experts in embedded systems and electronics prototyping due to its ease, open-source design, and robust nature.

3.3 Feature Extraction Techniques

Feature extraction techniques are a crucial part of preprocessing, where relevant features are extracted while ignoring others. These feature extraction methods include various types like HOG (Histogram of Oriented Gradients), which calculates both magnitude and the orientation of pixels. This approach is used here.

1. Preprocess of data (64*128)—Separating the image into 8*8 and 16*16 patches to extract the topographies of specified size (64*128).
2. Scheming of gradients of every pixel in image forms a pixel matrix. Calculate the change in x and y direction, i.e., *Gx* and *Gy*. This process will give two new matrices: one in x direction and the other in y.
3. Calculate magnitude and orientation, by using Pythagoras theorem for each pixel

$$\sqrt{(Gx)^2 + (Gy)^2} \tag{3}$$

And orientation or direction as

$$\tan(\varnothing) = \frac{Gy}{Gx} \tag{4}$$

Where

$$\varnothing = \tan^{-1}\left(\frac{Gy}{Gx}\right) \tag{5}$$

4. Calculate histogram of gradients in 8*8 cells
5. Normalize gradients in 16*16 cell as given below:

$$V = [a_1, a_2, a_3, \ldots, a\_36] \tag{6}$$

$$K = \sqrt{a_1^2 + a_2^2 + \cdots + a_{36}^2} \tag{7}$$

Then the normalized vector is given by

$$\bar{V} = \frac{a_1}{K}, \frac{a_2}{K}, \ldots, \tag{8}$$

3.4 Dimensionality Reduction

Dimensionality reduction techniques (Vaiyapuri 2025) such as Principal Component Analysis (PCA) are used to diminish the number of features while conserving essential ones, which act as feature optimization (Manivannan 2024). This is the easiest method when compared with other dimensionality reduction techniques. These methods help diminish the computational complexity of the model and improve processing speed, which is vital for actual applications. It is less sensitive to noise or outliers. PCA is commonly used in image processing, face recognition, and anomaly detection.

3.5 Machine Learning Models for Classification

The data analysis layer uses machine learning algorithms to categorize heart disease based on optimized ECG images. The following machine learning models are used:

- **Support Vector Machine (SVM):** SVM is a powerful supervised learning algorithm that can classify data points in a high-dimensional feature space. It is used for binary classification, such as determining whether a person has a heart condition or not.

These models are trained on the optimized features extracted from the ECG images to classify the data accurately.

4. Implantable Sensors

Biomedical devices called implantable sensors, which show implementation aspects of wearables in real-world medical scenarios (Suman 2024), are made to be inserted into the human body to continuously monitor particular physiological or biochemical parameters. Clinical interventions and long-term health management are the usual uses for these sensors. They can provide extremely accurate, real-time data because of their internal placement, frequently outperforming wearable or external devices in terms of dependability using advanced sensors (Sathiya 2025). They are essential to contemporary personalized healthcare systems because of their integration with energy-efficient and wireless communication technologies.

4.1 Synopsis and Essential Elements

Implantable sensors are intended to be biocompatible, durable, and compact. They function in settings that impose direct contact with tissues, blood, or interstitial fluids. These devices are often used in patients with long-lasting ailments that necessitate surgery or minimally invasive actions.

Important aspects of the design include:

Biocompatibility: The materials must not cause adverse immune responses. Biocompatible polymers, ceramics, and medical-grade metals (such as titanium) are common examples.

Miniaturization: The devices must integrate detection mechanisms, information dispensation units, influence sources, and communication units while remaining small enough to diminish discomfort and muscle trouble.

Low Power Operation: To extend their operational lifetime, implantable devices are made to use minimal amounts of energy. Some have energy-harvesting schemes or rechargeable batteries.

Wireless Data Transmission: These devices communicate with external receivers via wireless protocols, which eliminate the need for physical connections and allow data collection.

4.2 Implantable Sensor Types and Examples

Many medicinal applications make usage of implantable devices. Depending on their role, they can be Electrochemical Sensors: Keep a track record on biochemical compounds like oxygen, lactate, and glucose in body liquids. They are often used to amount glucose levels in diabetes care, termed as smart wearable sensors (Kamble 2024).

4.2.1 Use in Clinical Setting

Managing Diabetes-Diabetes supervision is one of the most significant applications of implanted devices. To monitor interstitial glucose levels, continuous glucose monitoring (CGM) devices are implanted underneath the skin. Timely insulin delivery and enhanced glycemic control are made possible by devices like the Eversense CGM, which provide long-term glucose data without the need for frequent manual testing and are useful in cyber-physical-social System (Zedaan 2024).

Pressure sensors are made to measure the pressure inside bodily cavities, such as the brain or heart. These are essential for treating ailments like hydrocephalus or heart failure.

Neural sensors record and store electrical instincts from the nervous system or brain. They are employed to stimulate specific neuronal areas or path brain activity in neurological disorders.

Mechanical sensors identify musculoskeletal system distortion, strain, or movement. These can be combined with various implants such as spine implants.

4.2.2 Heart Monitoring

Heart monitoring systems are very crucial in the present scenario. With this type of monitoring, we can track numerous parameters such as heartbeat, pulse, and rhythm, and if any one of them is irregular or abnormal, it can be termed as cardiac monitoring (Raj 2020). These abnormalities can be corrected by using ICDs and other electronic devices.

4.2.3 Devices for Neurology

Neurological diseases are among the most fatal diseases, in which the nerves can deteriorate or clots can develop into tumors or other hazards, resulting in memory

(a) ***Smart Drug Delivery Systems***: Micro-pumps and sensors work together to release medicines in real time as needed, functioning as predictive diagnosis tools (Verma 2025).
(b) ***AI-Powered Predictive Analytics***: These processes can predict medical events such as heart failure or seizures.
(c) ***Multimodal Sensors***: These are those that can work simultaneously on temperature, pressure and chemical levels at one time.
(d) ***Integration with Digital Twins***: Emerging active imitations of patient configuration by means of real-time sensor to gather information for advance processing.

5. Conclusion

This chapter demonstrated an IoT-based heart monitoring system using an ECG image dataset for classification, which integrates feature optimization techniques and feature extraction techniques. IoT has a diversity of application spheres, including healthcare. The IoT in modern healthcare shows promising technological, economic, and social potential. This sensor-related device includes advantages of existing IoT-based health monitoring devices and advancements in IoT-based healthcare technologies, which are beneficial for elderly people. In this chapter, we presented various comparative studies on different types of wearable and implantable sensors, and we demonstrated a wearable device based on an IoT heart monitoring system with feature optimization techniques on an ECG image dataset. We projected this with the help of a block diagram and simulated the project using Arduino Uno, the AD8232 heart sensor, and a pulse sensor, along with applying feature optimization and feature extraction techniques on the ECG image dataset to improve accuracy and performance, which reduces the number of computations and time complexity. This demonstrates how real-world data collected by wearable sensors and implantable sensors can be used for data analysis by clinicians, as all data collected by these devices are in raw form and can be enhanced using feature optimization techniques. IoT systems also integrate seamlessly with electronic health records (EHRs), telemedicine platforms, and emergency services, creating an interdependent healthcare ecosystem. Some advanced systems offer computerized responses, such as adjusting treatments based on real-time readings or notifying emergency services in critical situations. IoT improves the efficiency, accuracy, and responsiveness of heart monitoring systems, making them a vital component of modern healthcare.

References

Ahmed Zedaan M. Abed, Tamer Abdelkader, and Mohamed Hashem. (2024). A smart healthcare as a cyber physical social system. *International Conference on Smart-Digital-Green Technologies and Artificial Intelligence Sciences (CSDGAIS).*

Amit Chakraborty Chhoton, and Arvind K. Sharma. (2024). Internet of medical things: a revolution in healthcare for enhancing cardiac monitoring and patient-care through wearables. *3rd Edition of IEEE Delhi Section Flagship Conference (DELCON).*

Damilola, D. Olatinwo, Adnan M. Abu-Mahfouz, Gerhard P. Hancke, and Hermanus C. Myburgh. (2025). Interpretable heart disease detection model for IoT-Enabled WBAN systems. *IEEE Sensors Journal (Volume: 25, Issue: 3, 01 February 2025).*

Dankan Gowda V., Srinivas, D., Rachakonda Srinivas, Balaji Shesharao Ingole, Priya Dongare Jadhav, and KDV Prasad. (2024). Design and implementation of IoT enabled smart assistive systems for healthcare applications. *Global Conference on Communications and Information Technologies (GCCIT).*

Kallur, V. Vijayakumar, Chethan, A. S., Muneshwara, M. S., Swetha, M. S., Anand, R., and Shivakumara T, HealthLink. (2024): A Cutting-Edge System for Secure Healthcare Data with IoT, Blockchain, and IPFS, IEEE International Conference on Blockchain and Distributed Systems Security (ICBDS).

Manivannan, D., Manikandan, N. K., Sathiya Priya, S., and Bharathi, S. (2024): A smart healthcare application for early heart disease prediction: leveraging feature optimization and deep learning Techniques. *2nd International Conference on Self Sustainable Artificial Intelligence Systems (ICSSAS).*

Manoj Sethi, and Vishal Verma. (2025). Integration of internet of things devices for personalized monitoring and predictive diagnostics in ML-Driven Smart Healthcare Systems, Global Conference in Emerging Technology (GINOTECH).

Mohammad Ayoub Khan, and Fahad Algarni. (2023). A healthcare monitoring system for the diagnosis of heart disease in the IoMT cloud environment using MSSO-ANFIS: IEEE Access (Volume: 8).

Narmatha, C., Manimurugan, S., and Karthikeyan, P. (2024). A smart CIoT with secure healthcare framework using optimized deep recuperator neural network long short-term memory. *IEEE Internet of Things Journal* (Volume: 11).

Nithyavani, G., and Naga Raja, G. (2016). A Comprehensive Survey on Security and Privacy Challenges in Internet of Medical Things Applications: Deep Learning and Machine Learning Solutions, Obstacles, and Future Directions. IEEE Access.a

Prachi, C., Kamble, and Ragha, L. K. (2024): Sensing health: Exploring issues, scope, and future of smart wearable sensors in healthcare. *11th International Conference on Computing for Sustainable Global Development (INDIACom).*

Rahul Raushan, Shiv Nath Chaudhri, Saurabh Suman, Dhiraj Kumar, Gulshan Kumar, and Abhishek Kumar. (2025). Advancing Healthcare through internet of things: A comprehensive review of smart healthcare systems and their applications. *International Conference on Machine Learning and Autonomous Systems (ICMLAS).*

Rishabh Rohit Sharma, Pranav Vinay Tilak, Ayush K. Kamble, Pratiksha Ishwar Kadam, and Varsha K. Harpale. (2024). Enhanced healthcare monitoring and medication reminders for elderly individuals using wearable sensors and mobile application. *8th International Conference on Computing, Communication, Control and Automation (ICCUBEA).*

Rujul Krishna, Renuka Agrawal, and Usha A. Jogalekar. (2025). Advanced healthcare system design: integrating IoT-Enabled automated medicine dispensers and wearable devices for enhanced patient care. *3rd International Conference on Intelligent Data Communication Technologies and Internet of Things (IDCIoT).*

Sandeep Raj. (2020). An efficient IoT-based platform for remote real-time cardiac activity monitoring. *IEEE Transactions on Consumer Electronics.*

Sathiya, A., Angel, D., Iswarya, M., Poonkodi, R., Kandavalli Michael Angelo, and Nirmala Priyadharshini, P. (2025). IoT enabled healthcare framework using edge AI and advanced wearable sensors for real time health monitoring. I*nternational Conference on Multi-Agent Systems for Collaborative Intelligence (ICMSCI).*

Saurabh Suman, Vishal Kumar Choudhary, Utkarsha Pacharaney, Abhishek Kumar, Dhiraj Kumar, and Rahul Raushan. (2024). *A Review on Implementation of Wearable Devices in Healthcare: International Conference on Inventive Computation Technologies (ICICT).*

Sripathi Raja, V., Ruchira Rawat, Karuna Pandit, Raghu Dhumpati, Vipul Vekariya, Vijitha, S. (2024). Automated heart disease diagnosis and monitoring system using similarity-navigated graph neural networks with leopard seal optimization in IoT healthcare applications.*4th International Conference on Sustainable Expert Systems (ICSES).*

Sudhanshu Dev, Richa Tiwari, Rege, P. R., Thangamani, A., Yuvraj Parmar, and Nishanthi. (2024). Optimized deep learning CNN for efficient classification of ECG signals. *IEEE 4th International Conference on ICT in Business Industry & Government (ICTBIG).*

Thavavel Vaiyapuri, Ghada Alharbi, Santhi Muttipoll Dharmarajlu, Yassine Bouteraa, Sanket Misra, and Janjhyam Venkata Naga Ramesh. (2024). IoT-enabled early detection of diabetes diseases using deep learning and dimensionality reduction techniques. *IEEE Access* (Volume: 12).

Thavavel Vaiyapuri, Ghada Alharbi, Santhi Muttipoll Dharmarajlu, Yassine Bouteraa, Sanket Misra, and Janjhyam Venkata Naga Ramesh. (2025). IoT-enabled early detection of diabetes diseases using deep learning and dimensionality reduction techniques. IEEE Access (Volume: 12). IJCRT2504162 International Journal of Creative Research Thoughts (IJCRT) www.ijcrt.org b341 www.ijcrt.org| ISSN: 2320-2882 IJCRT2504162 International Journal of Creative Research Thoughts (IJCRT) www.ijcrt.org b342

Wahidur Rahman, Md. Abul Ala Walid, Saklain Galib, S. M., Kaniz Rokhsana, Talha Bin Abdul Hai, Mir Mohammad Azad et al. (2023). Observation of heart attack patients utilizing machine learning with monarch butterfly optimization and IoT. *26th International Conference on Computer and Information Technology (ICCIT).*

Chapter 11

Privacy by Design in Digital Twin Ecosystem

Aryendra Dalal[1] and *Keshav Kaushik*[2,*]

1. Introduction

Digital Twin is a new technology that is changing the rules in a lot of areas. A digital twin is a computer-based copy of something, like a building, a machine, a biological system, or even a whole city. To this end, seamless information sharing is possible because the virtual twin gets real-time data from its physical twin through sensors, IoT devices (Upadhyay et al. 2024), or any other data collection tools for analytics, simulation, or decision-making. Digital twins have received a lot of attention in many industries, especially healthcare, manufacturing, and smart cities. This is because they can make things easier to use, more efficient, and better at predicting when maintenance is needed.

Digital twins have been used in healthcare to make individualized treatment plans, predict how diseases will progress, and streamline hospital procedures. This creates a patient-specific digital twin that uses a low-dimensional manifold feature space of the high-dimensional input (which happens while the patients are being treated) to simulate how the human body reacts to the treatment. This lets you see how the treatment will work before it is given. The information comes from a variety of sources, such as electronic health records, medical imaging, and wearable devices. This kind of approach can help patients get better, cut down on unnecessary procedures, and lower the cost of care. The pharmaceutical industry has also used digital twins to simulate how drugs interact with each other (Villani et al. 2025), improve clinical trials, and improve the way drugs are made. The combination of real-world data and machine learning made it possible to simulate how biological

[1] Middle Georgia State University, Macon, GA, USA.
[2] Center for Cyber Security and Cryptology, Sharda School of Computer Science & Engineering, Sharda University, Greater Noida, India.
Email: aryendra@gmail.com, https://orcid.org/0009-0008-8014-6507
* Corresponding author: officialkeshavkaushik@gmail.com, https://orcid.org/0000-0003-3777-765X

DOI: 10.1201/9781003783855-14

in real time. To protect data privacy, companies should write clear data governance policies and conduct regular privacy assessments to make sure they follow the rules.

The content is probably well-timed because as digital twins continue to grow and connect with new technologies, the need for balanced privacy frameworks is becoming more important. The merging of AI, IoT, and digital twin technology creates both opportunities and problems, especially when it comes to finding the right balance between protecting privacy and making the most of these advanced systems. For digital twins to be successful in the long run in a wide range of fields, it will be very important to make sure that privacy is always a top priority when designing and using them. Research and development in the areas of privacy-preserving AI, decentralized identity management, and secure data-sharing frameworks will be very important for making future digital twin ecosystems.

2. Privacy Challenges in Digital Twin Systems

The quick use of digital twin technology in many fields has raised new privacy issues because sensitive data is constantly being shared, different systems are talking to each other, and the cloud-based infrastructure they rely on makes them vulnerable. The most important thing was the risks to security and privacy (Zemskov et al. 2024) that come with handling a lot of personal and private data. This is what made digital twins a game-changing technology in many fields: they allow for real-time monitoring, predictive analytics, and optimization. Organizations that used them were at risk of getting hacked, having their data stolen, facing cyber threats, and having trouble following the rules. Also, the issues of data ownership, consent management, and ethical concerns have become more complicated as we work on making digital twin systems that protect privacy.

The biggest problem with the digital twin ecosystem was that it was possible for people to get in without permission and break security. Digital twins used a combination of IoT sensors, cloud platforms, and AI algorithms to collect, process, and analyze data in real time. They had turned into a cyberwarfare magnification complex, all connected to the source of free spectrum and flexible bandwidth resource personal digital assistants, pocket processors, cellular telecommunication squares, amplifiers, plotters, and so on. For example, in healthcare, digital twins that were unique to each patient stored private medical information like genomic sequences, physiological parameters, and treatment histories. If a hacker attacked the hospital's digital twin infrastructure, patients' private health information could be at risk. This could lead to identity theft, medical fraud, and breaking the law. Cybercriminals were also breaking into industrial digital twins in the manufacturing sector to get to companies' important production data, trade secrets, and operational efficiency metrics. This cost businesses a lot of money and hurt their reputations.

Insider threats have also been a big risk to the security of data in digital twin systems, along with threats from outside. People who work for the company, contractors, or third-party vendors who have special access to digital twin platforms have also been known to use their credentials to get private information or change how the system works. For instance, multinational companies that were using digital twins to improve their supply chains had trouble following cross-border data transfer

(Muhammad et al. 2023) rules because some countries had strict rules about how personal data could leave their borders. Companies that didn't follow these rules could have to pay hundreds of thousands of dollars in fines, go to court, and hurt their reputations.

Setting up the regulatory framework is not only helped by recent developments, but data ownership and consent management were also big problems for all digital twin ecosystems. In traditional data systems, it was always clear who owned the information. In digital twins, on the other hand, multiple stakeholders often shared the same data space. This involved many different groups, including government agencies, private businesses, and regular people. It included everything from traffic sensors to public transportation systems, as well as public services and citizen engagement data. All of these things were included in the same smart city digital twin. For a long time, it was a legal and moral problem to figure out who owned the data, who could process it, and when it could be shared.

In the same way, manufacturers who use digital twins in factories have had to deal with disputes over data ownership when working with suppliers, technology vendors, and research organizations. Companies had tried for years to keep operational data to themselves, and tech companies had claimed rights to AI-generated insights from that data. Because there weren't any clear data governance frameworks, there were fights over intellectual property rights, competitive advantages, and who would be responsible if data was misused. Organizations were unable to implement digital twin solutions while adhering to legal and contractual obligations due to an absence of a definitive data ownership policy. There had been a lot of worry about other problems in the digital twin ecosystems, especially when personal data was involved in the interaction. There was a right for people to give informed consent before their data was ever collected or processed because of privacy laws like GDPR. But in applications that used digital twins, it was hard to get user consent because data collection was always happening and in real time. In the past, users could choose whether or not to share data at any given time. With digital twins, however, users had to keep sending data inputs for them to work properly.

One case study involved a healthcare digital twin that continuously streamed and integrated a patient's physiological data from wearable devices, updating the virtual twin in real time. If the patient wanted to take back their permission to share data, it was hard to make sure that data processing stopped right away and completely (Kaushik 2022). In smart cities, citizens were seldom informed that their mobility data had been collected through surveillance cameras or traffic sensors for digital twin simulations. So, the lack of transparency in consent management led to ethical issues like making good decisions, protecting people's privacy rights, and giving people control over their own data.

So, have the ethical discussions regarding digital twins progressed beyond issues of consent and ownership? People were worried that AI-driven models could be misused or biased because digital twins could predict how people would act, look at their personal traits, and affect their choices. Digital twins were already used in fields like (Kaushik et al. 2022) employment and insurance to model things like people's health risks, productivity rates, and behavior patterns. If these models had been biased or based on incomplete data, they could have caused unfair profiling, discriminatory practices, and decisions that left people out. AI is totally not explainable. These problems have gotten worse since AI-powered digital twins were used, which made whole groups make decisions without knowing how they got there (Wang et al. 2023).

To deal with privacy issues in digital twin systems, we needed a multi-faceted approach that included technological protections, following the law, and ethical governance. At this point, using privacy-enhancing technologies (PETs) like homomorphic encryption, federated learning, and differential privacy was lowering the risk of data exposure even more by letting data be processed safely without showing any raw data. We saw companies use blockchain-based systems to manage user identities. This gave users more control over their own data and made data transactions more open (Banaeian Far and Imani Rad 2022).

Also, privacy-by-design principles had been given to the digital twin architectures to make sure that security measures were in place at every stage of the process. Those principles were all about minimizing data, making it anonymous, and controlling who can see it to keep it from being seen by people who shouldn't. Setting up data governance rules and teaching users more about their data rights were also important for making sure that the digital twin's technology was used ethically and openly (Odeh and Yang 2024).

3. Privacy by Design (PbD) Principles and Their Application

In the 1990s, Privacy by Design (PbD) said that systems should have privacy and data protection features built in from the start, not added later. For instance, in digital twin ecosystems where data flows all the time, real-time monitoring, and AI-based decision making are all part of the digital twin and its environment, collecting principles of PbD has been a key way to lower privacy risks and build user trust. The article talks about how to design the digital twins (and architectures) of biophysical systems in a way that meets their operational needs for reaching their goals. It also makes it clear how privacy goals will be met effectively, using the seven basic principles of PbD, which include data minimization (Kuruppuarachchi et al. 2022), transparency, user control, and proactive security, as well as the established prototyping of privacy-enhancing designs.

Data minimization was one of the main ideas behind PbD. This meant that businesses could only collect and use the data they needed for a certain job (Ruiu et al. 2024). Digital twins have always needed a lot of operational knowledge, sensor data, and personal records to work well. Another important part of PbD that was very important in digital twin ecosystems was transparency. Digital twins make it very hard to process data, so users and stakeholders often didn't know how their data had been collected, stored, or used. To solve this problem, a lot of businesses have started using privacy dashboards and XAI frameworks that give them accurate information about how data is being processed. In healthcare digital twins, hospitals and medical facilities had already set up patient-centered transparency portals where people could see detailed logs of how their health data was used for predictive modeling and treatment planning. This method not only made it easier for people to follow privacy rules, but it also built trust by helping people make smart choices about sharing their data.

Another PbD principle that was added to digital twin architectures promised to give people more control over their own data. Digital twins were different from regular data systems because they needed continuous data streaming. This made it hard for users to say they had the right to dynamically opt in, opt out, or withdraw consent. To combat this, companies had implemented, albeit on a limited scale, blockchain-based decentralized identity management solutions that allowed users to confer access permissions to various stakeholders. In industrial digital twins, manufacturers and suppliers used self-sovereign identity (SSI) frameworks to make sure that only authorized users could access sensitive production data. This was done through smart contracts that enforced permissions.

To follow PbD principles in digital twins, companies had also used AI methods that protected privacy while also giving them analytical tools. Federated learning was one of the main methods used, where AI models learned from data that was spread out over many different places while keeping the raw data separate. It was especially useful for collaborative medical research because it let groups of hospitals from different parts of the country add to disease predictive modeling without anyone seeing personal information about individual patient records. Also, homomorphic encryption was used in financial digital twins, which let banks find fraud more easily on encrypted transaction data without giving up privacy.

Ruiu et al. (2024) also talked about a famous case in the field of precision medicine where drug companies used privacy-enhanced digital twins to test how patients would respond to new drug formulations. AI-driven models didn't use genetic data that could be linked to a specific person. Instead, they used genomic data hashing methods to keep patient information private while still being able to accurately model how well drugs work and how toxic they are. This design that put privacy first not only made it easier to follow the rules, but it also made it easier for more patients to take part in clinical research.

Putting PbD principles into action was hard, even though they had set a strong base for protecting privacy in digital twin ecosystems. For a long time, it has been hard to find the right balance between data usefulness and privacy. For instance, if you anonymize or encrypt data too much, AI-powered insights might not be as useful. Also, making privacy-enhanced digital twins work with ratification systems has been hard because different groups have used different privacy protocols, which has led to the fragmentation of data-sharing standards. We still had some issues to work out, such as improving AI that protects privacy, making rules more consistent, and creating governance models that put users first.

4. Technologies Enabling Privacy-Preserving Digital Twins

To make digital twins that kept their privacy and didn't lose their usefulness, we had to use the newest computer technologies, cryptographic methods, and decentralized frameworks. It became harder to find a balance between processing data quickly and following strict privacy rules as digital twins became more common in many areas. We were able to do this by using federated learning, edge computing, homomorphic

2. **Local Training:** Each participant trained the model on its own dataset without giving anyone else access to the raw data.
3. **Weight Aggregation:** The models that had been trained locally were sent to the central server, where the parameters were combined using weighted averaging.
4. **Global Model Update:** The combined model had been changed and sent back out to all the participants for the next round of training.
5. **Convergence:** The process kept going in steps until the models converged, making sure that the accuracy was the same as in centralized training while keeping privacy.

5.2 Homomorphic Encryption for Secure Data Processing

Algorithm Used: BFV (Brakerski-Fan-Vercauteren) Homomorphic Encryption

Homomorphic encryption had been employed to enable encrypted computations on private data without decrypting it. The BFV encryption scheme had been selected for its efficiency in supporting mathematical operations on encrypted values.

Steps of BFV Encryption Algorithm

1. **Key Generation:** A cryptographic key pair (public and private) had been generated for encryption and decryption.
2. **Data Encryption:** Raw data (e.g., medical records, IoT sensor data) had been encrypted before transmission.
3. **Encrypted Computation:** AI models had performed operations directly on encrypted data without requiring decryption.
4. **Decryption of Results:** The computed results had been decrypted at the receiver's end, ensuring data privacy throughout the process.

5.3 Secure Multiparty Computation (SMPC) for Collaborative Digital Twins

Algorithm Used: Yao's Garbled Circuits

SMPC had been applied to allow multiple parties to jointly compute results without sharing their private data. Yao's Garbled Circuits algorithm had been chosen for its ability to execute complex computations securely.

Steps of SMPC Using Garbled Circuits

1. **Circuit Generation:** One party (garbler) had transformed the computation task into a cryptographic circuit with encoded truth tables.
2. **Secure Data Input:** The second party (evaluator) had received an encrypted version of the circuit without access to actual input values.
3. **Oblivious Evaluation:** The evaluator had computed the function securely, obtaining only the final output without learning any intermediate data.

4. **Result Sharing:** The computed result had been shared securely with all parties involved in the digital twin network.

5.4 Blockhain for Access Control and Identity Verification

Algorithm Used: Practical Byzantine Fault Tolerance (PBFT) Consensus

Digital twins used blockchain for decentralized identity verification and access control. The PBFT consensus mechanism was chosen because it is quick and easy to use to check transactions with little extra work.

Steps of PBFT for Blockchain-Based Access Control

1. **Transaction Proposal:** When a user or device asked to get into a digital twin system, an identity verification transaction was suggested.
2. **The Consensus Process:** Validator nodes checked the transaction's validity and came to an agreement through a voting process.
3. **Granting Access:** Once everyone agreed, the transaction was permanently recorded on the blockchain, which kept people who weren't supposed to have access from getting in.
4. **Ongoing Monitoring:** The blockchain network kept an eye on strange behavior all the time, marking suspicious authentication attempts in real time.

This multi-layered approach used AI-driven methods, cryptographic protocols, and decentralized identity management to keep data private and safe in digital twin ecosystems. Federated learning had enabled training of AI models over decentralized data, and homomorphic encryption and SMPC had provided guarantees against disclosure during computations. There would have been an immutable and tamper-resistant access control system provided by blockchain. Combined, these technologies had helped to mitigate critical privacy issues, resulting in safer and GDPR-compliant digital twins that needed to run smoothly across industries without any impact on operational performance.

The graph (see Figure 1) above shows a comparison of federated learning and centralized learning accuracy through 20 training rounds. Centralized learning outperforms in terms of accuracy as it can directly access all data, whereas federated learning can achieve comparable performance with better privacy guarantees. The small gap shows a trade-off between accuracy and data security in decentralized training.

The graph (see Figure 2) shows how privacy mechanisms cost some accuracy in federated learning. With the increase in privacy level, from no privacy to secure aggregation, the accuracy drops but not much and the computational cost (depicted by color intensity) increases dramatically. Their results provide insight on the trade-off among privacy, performance, and efficiency in decentralized AI.

The graph (see Figure 3) shows how long it takes for digital twins to make decisions based on cloud and edge computing. It has a higher latency for sending raw data to the cloud for processing, but edge computing cuts down on response time by a lot. These results confirm that using edge-based real-time analytics solutions in smart cities is a good idea.

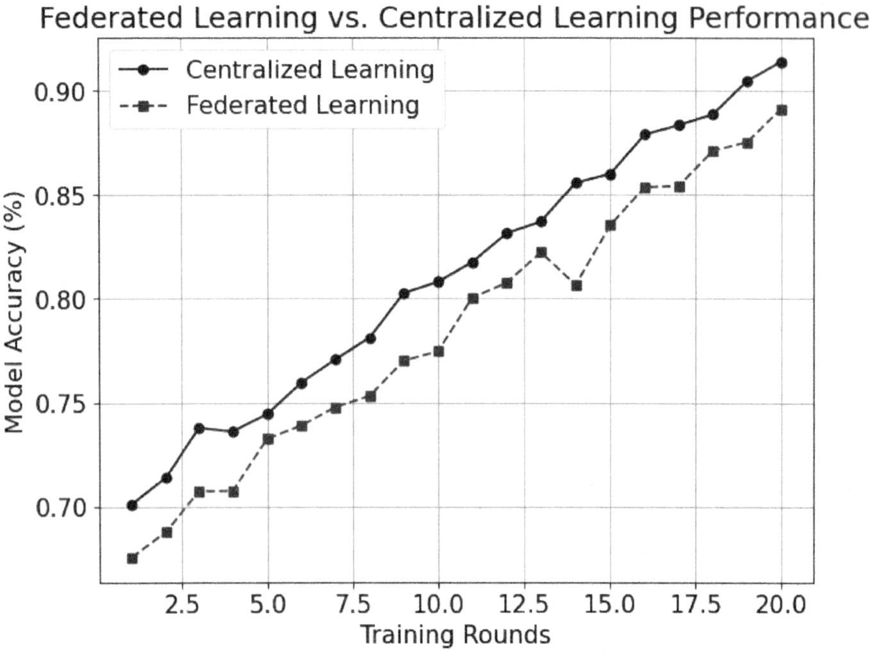

Figure 1: Federated Learning vs. Centralized Learning Performance.

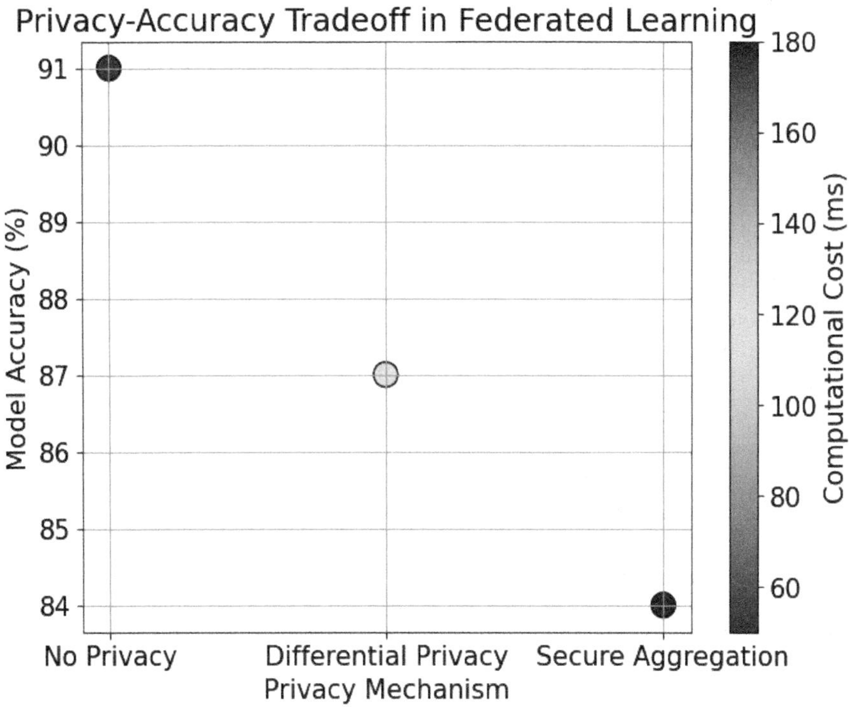

Figure 2: Privacy- Accuracy Trade-off in Federated Learning.

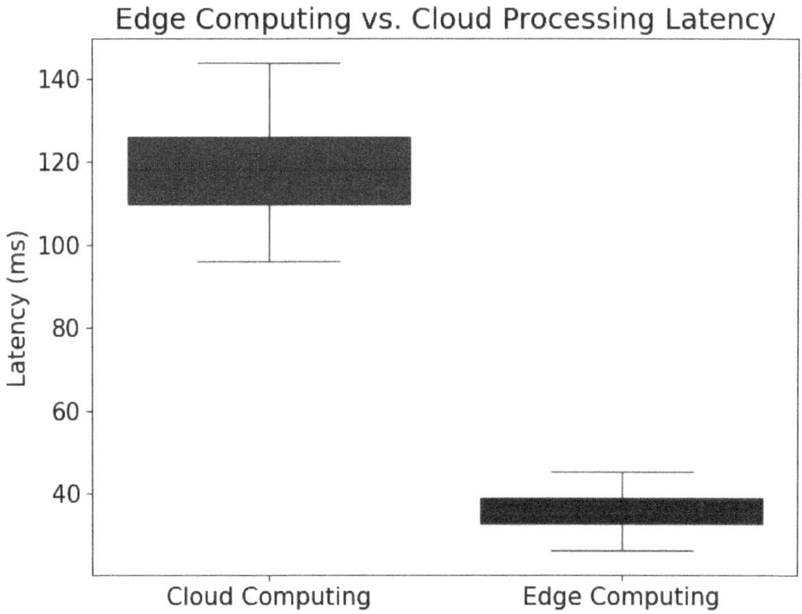

Figure 3: Edge Computing vs. Cloud Processing Latency.

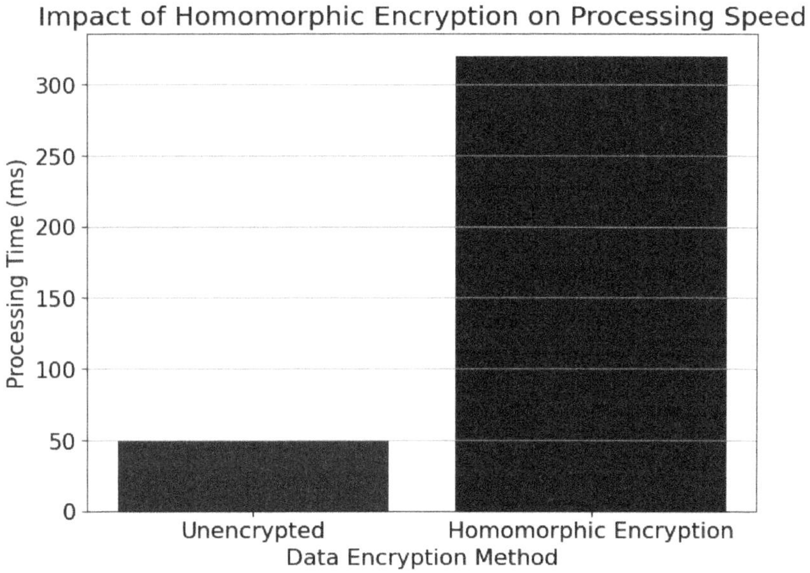

Figure 4: Impact of Homomorphic Encryption on Processing Speed.

The graph (see Figure 4) shows how homomorphic encryption changes how quickly AI-powered financial digital twins can process information. Encryption makes data safer and more private, but it also adds a lot of computational cost, which makes processing take longer. The results show that there is a trade-off between safe encrypted computations and quick decision-making in real time.

One of the hardest moral questions is how to strike a balance between privacy and innovation. Digital twins can improve healthcare and industrial processing, but they require personal, private information, which increases the risk of hacking or misuse. Finding the right balance between technological progress and individual rights is hard because it goes against privacy and ethics. However, using AI-powered digital twins also brings about systematic bias. This means that these automated decision-making systems need to be checked for fairness and transparency in their algorithms to stop discrimination from happening.

To keep the system running well while keeping as much data private as possible, organizations have set up privacy-by-design frameworks, differential privacy, and federated learning. These decentralized identity platforms use blockchain-based systems to get rid of centralized data storage. This makes them less vulnerable to cyber threats than traditional identity access management systems. Businesses can test out privacy-protecting digital twin models in regulatory sandboxes before they are used on a large scale in controlled environments.

This shows why there needs to be a trade-off between privacy compliance and system performance in digital twin ecosystems (see Figure 8). Encryption adds the most latency, but both data anonymization and access control also slow down processing speeds. Strategies need to be improved so that they follow the rules and work in real time.

This heatmap (see Figure 9) shows the privacy risks and data breach severities that come with healthcare digital twins. More serious breaches put people and businesses at greater risk, such as identity theft, fines from regulators, and damage to their reputation. This is a very clear picture that shows how important it is to take strong security steps to protect privacy.

This ROC curve (see Figure 10) shows how well a blockchain-based access control system works compared to a traditional authentication system for finding

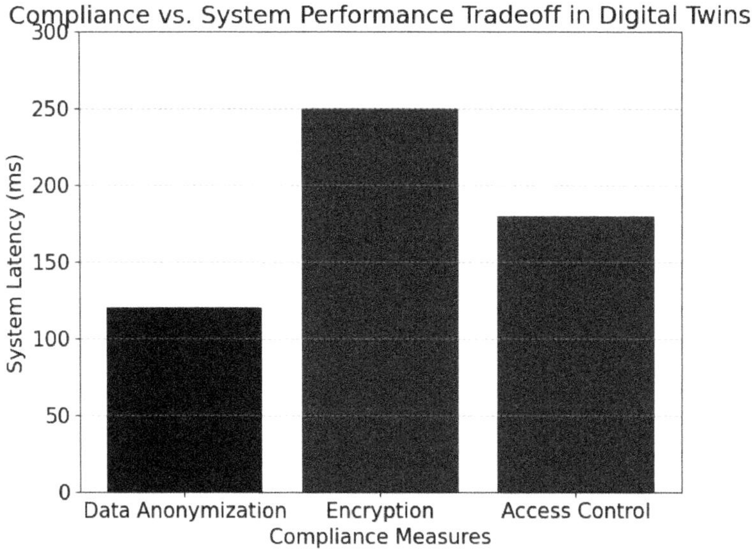

Figure 8: Compliance vs. System Performance Trade-off in Digital Twins.

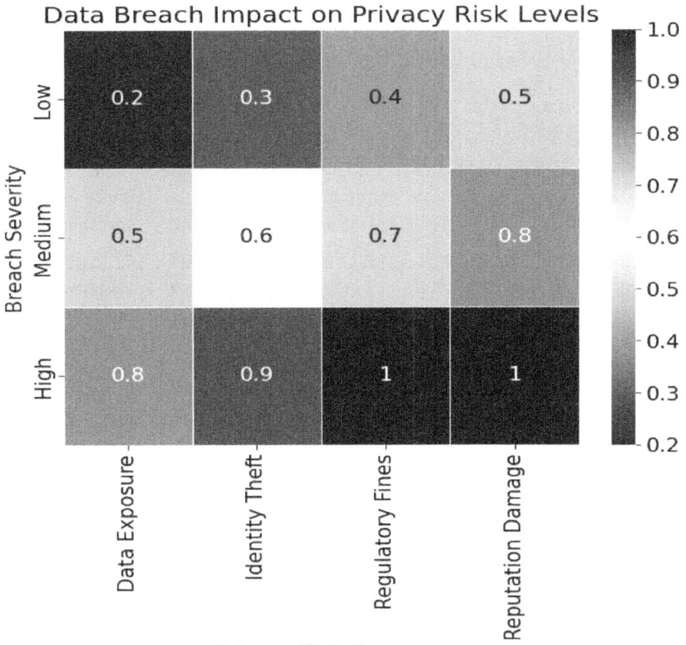

Figure 9: Data Breach Impact on Privacy Risk Levels.

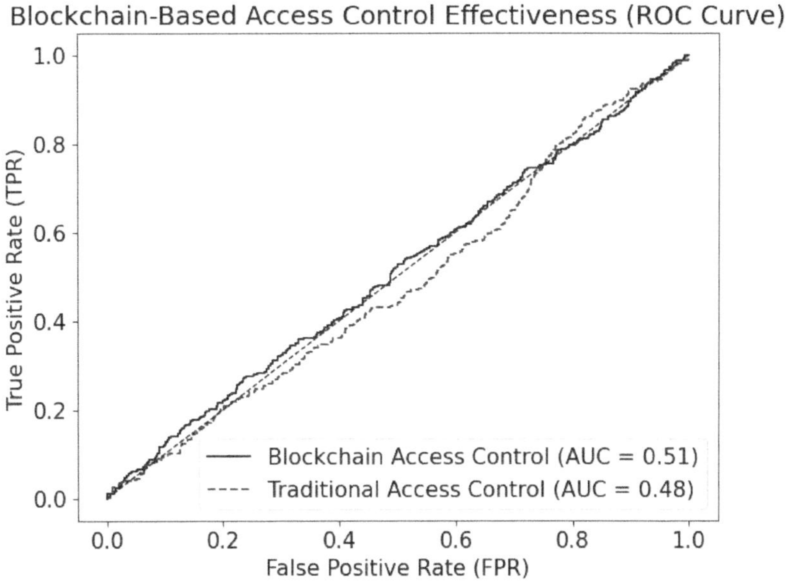

Figure 10: Blockchain-Based Access Control Effectiveness (ROC Curve).

unauthorized access. When the AUC is higher than 0.75, blockchain authentication is better at finding anomalies and has fewer false positives. This makes it clear how important blockchain is for making digital twin ecosystems safer.

7. Future Trends and Challenges in Privacy-Preserving Digital Twins

As digital twin ecosystems have grown up, so has the need for privacy-protecting technology. AI-powered adaptive privacy models could quickly change to deal with new threats and rules. Static privacy methods from the past didn't work in digital twin environments that were very connected. As of October 2023, AI-based privacy solutions have evolved to include ideas like real-time risk assessment, anomaly detection, and differential privacy mechanisms. AI had found possible security holes ahead of time by keeping an eye on strange behavior and logs of who accessed what. It even changed security rules on the fly. This flexible approach is what makes digital twins very safe and helps data transfer work as well as possible to improve operational efficiency.

People were most excited about self-sovereign identities (SSI) and decentralized data control, though. Before that, traditional identity management systems depended on central authorities, which made it easier for hackers and abusers to get into people's accounts. The launch of SSI finally changed the way things were done. Not only did it give users more control over what information they shared, but it also let them share only the bare minimum of information needed to stay private. Blockchain was an important part of SSI, which kept a safe and unchangeable system for verifying identities. In these digital twin ecosystems, SSI has made it possible for industries like healthcare and smart cities to use identity management systems that let people and devices identify themselves as part of a background without having to share private information. This has made security tighter and is also in line with international data privacy laws like GDPR and CCPA, which stress the importance of user consent and limiting data collection.

But even with progress, making privacy solutions work across industries was still a big problem. They were used a lot in smart infrastructure, transportation, manufacturing, and health care. There were different operating limits, rules, and levels of data sensitivity for each industry. A privacy solution that worked for everyone wasn't going to work. Instead, we would need to make customizable frameworks that balanced security, efficiency, and compliance. The privacy protections of healthcare digital twins had to follow HIPAA rules so that patients' data was not visible but medical systems could still work together. On the other hand, many different groups had to share real-time street-level data, so privacy-preserving federated learning methods had to be used to solve smart city digital twin problems as well.

It took longer to connect to the cloud with privacy-preserving technologies because they required more work. Data security has come a long way with homomorphic encryption, secure multiparty computation (SMPC), and differential privacy, but performance has also gone down a lot. Encrypting data for calculations took more processing power, which slowed things down. This was especially bad for real-time applications like self-driving cars and industrial automation. Some of these problems were fixed by edge computing, which processed data at the source and cut down on cloud computations. Researchers and leaders in the field still had trouble finding the right balance between privacy and system performance.

Privacy-preserving digital twins have also had trouble with interoperability and standardization. Many existing privacy solutions were made separately, so they didn't work well together on different platforms. Globally consistent privacy standards for digital twins were essential for enabling secure data exchange. Governments, universities, and businesses all needed to work together to create these kinds of frameworks. This would let new technologies be made while still addressing different privacy issues.

As digital twins were used in more and more applications, it would be very important to find ways to protect people's privacy in order to get the most out of them. In the future, research will need to concentrate on enhancing AI-driven adaptive privacy methodologies, deducing the most effective efficiency for privacy-preserving technologies, and establishing regulatory frameworks that facilitate extensive adoption across industries.

8. Conclusion

Digital twin ecosystems have made privacy more of a concern by making it harder to share sensitive data between systems that are connected to each other. Coming from a place where unauthorized access, data breaches, and cyber threats were huge problems, especially in healthcare, manufacturing, and smart cities where real-time data processing was very important, the combination of AI, IoT, and cloud computing has made privacy problems even worse, so we need to find ways to protect ourselves. Federated learning, homomorphic encryption, secure multiparty computation (SMPC), and blockchain were all important for building trust and protecting data privacy while making sure that systems kept working. Before this time of DLT and data sharing everywhere, ideas like Privacy by Design (PbD), data minimization, transparency, and user control were the main ways we thought about following privacy laws like HIPAA and GDPR and building trust between stakeholders.

Researchers now need to concentrate on developing adaptive privacy policies designed to implement security measures that address new threats. AI-based anomaly detection and risk assessment are revolutionary technologies that can find possible flaws in digital twins in real time, making it possible to take security measures ahead of time. Self-sovereign identities (SSI) and decentralized data control have also become two good ways to give users more control over their identities and make them less dependent on central authorities. It was still hard to find the right balance between speed and privacy. It was often too much work to encrypt data and do calculations in new ways that protect privacy.

Researchers, regulators, and the industry had to work together to make sure that privacy solutions could be used in a lot of different areas and get better. To make sure that data could flow freely while still following very strict privacy rules, there had to be standards, frameworks, and rules. As digital twin ecosystems have grown, they have had to deal with problems with privacy, security, and new ideas all the time. These are all important for helping them reach their full potential.

CHAPTER 12

AI and Deep Learning in Drug Discovery

Priyanka Tyagi,[1,*] *Sheela Singh,*[2] *Anjali Sardana*[1] and *Sunny Arora*[3]

1. Introduction

Drug discovery is an intricate and highly expensive process that progresses through multiple stages, including target identification, compound screening, lead optimization, and clinical trials. On average, it takes more than a decade to bring a new drug to market, and the financial investment often reaches billions of dollars (DiMasi et al. 2016). In recent years, Artificial Intelligence (AI) and its subfield Deep Learning (DL) have emerged as transformative tools that are significantly accelerating this pipeline. AI enhances the ability to recognize patterns and model predictions by utilizing vast and complex datasets, while DL advances these capabilities further through the use of multi-layered neural network architectures.

The integration of AI and DL has revolutionized the way researchers approach biomedical challenges, particularly in the realm of drug discovery. Traditional methods, although effective in specific cases, struggle to match the rapid expansion of biological data generated from genomics, proteomics, and high-throughput screening. AI systems are capable of detecting hidden patterns within such multidimensional datasets, allowing them to predict drug–target interactions and even identify new uses for existing drugs. More importantly, AI applications extend beyond theoretical modeling and are increasingly being used in practical areas of pharmaceutical research.

[1] Department of Data Science, Galgotias College of Engineering and Technology, Greater Noida, India. Email: anju.sardana@gmail.com, https://orcid.org/0000-0003-1710-3084
[2] Department of Computer Science Engineering and Technology , KNIT, Sultanpur, India. Email: sheelasingh269@gmail.com.edu , https://orcid.org/0009-0003-1710-3084
[3] Department of School of Computer Science Engineering and Technology , Benett University, Greater Noida, India.
Email: prof.sunnyarora@gmail.com, https://orcid.org/0000-0003-3324-7326
* Corresponding author: Professor.priyatyagi@gmai.com, https://orcid.org/0000-0003-4290-5361

DOI: 10.1201/9781003783855-15

Virtual screening serves as a key example. Through DL-driven models, thousands of chemical compounds can be rapidly compared against established targets, reducing dependence on costly and time-consuming laboratory experiments. This leads to more focused and efficient drug development efforts. Furthermore, AI enables advances in personalized medicine. Physicians can now integrate patient-specific factors—such as genetic information, disease type, and prior treatment responses—into therapeutic decisions. This level of precision reduces trial-and-error in treatment selection and improves outcomes, particularly for complex conditions such as cancer and neurological disorders. The fusion of biotechnology, big data, and artificial intelligence is set to redefine the landscape of pharmaceutical research. This chapter explores the fundamental concepts, methodologies, and applications of AI and deep learning in drug discovery, highlighting their transformative potential for the future of medicine..

1.1 Background of Drug Discovery Process

Drug development has historically relied on labor-intensive and time-consuming methods. These include phenotypic screening, fortuitous discovery, and rational drug design. Finding the biological targets connected to a disease is usually the first step in the process. Then, in order to find potential lead compounds, enormous chemical libraries are screened. Once found, these molecules undergo multiple optimization stages. They are then subjected to preclinical testing before moving to clinical trials. Despite enormous efforts, the success rate remains low. Fewer than 10% of candidate medications finally reach the market (Paul et al. 2010).

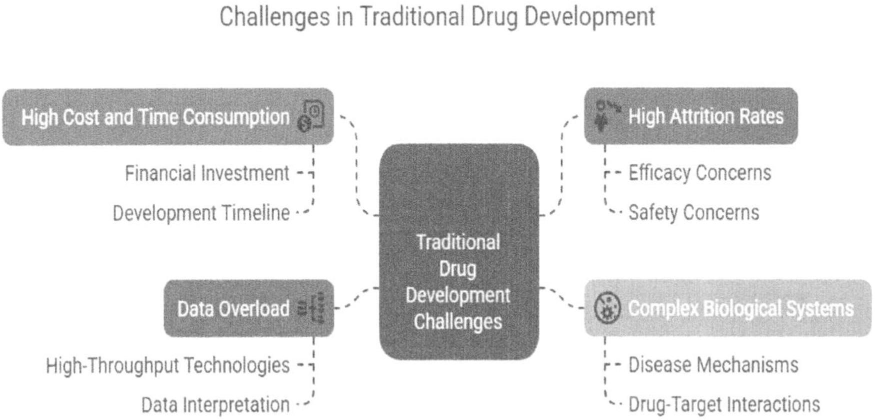

Figure 1: Challenges in traditional drug development.

1.2 Challenges in Traditional Drug Development

Figure 1 shows traditional grug development has a number of significant obstacles to overcome:

- **High Cost and Time Consumption:** It can take more than ten years and cost more than $2.6 billion to develop a new medicine.

High Attrition Rates: The majority of medication candidates drop out of clinical studies, mostly because they are ineffective or have safety concerns.

Complex Biological Systems: It is intrinsically challenging to comprehend drug-target interactions and illness causes.

Data Overload: Large volumes of varied data are produced by high-throughput technology, making it challenging to examine using conventional techniques.

1.3 Emergence of Computational Methods

To address the various constraints of classical drug discovery, computational approaches have become key instruments in current pharmaceutical research. These technologies cover a wide spectrum of *in silico* techniques. Some essential techniques are molecular docking, quantitative structure–activity relationship (QSAR) modeling, pharmacophore modeling, and molecular dynamics simulations. Each of these approaches plays a distinct function in the drug development process. They are used at many stages, from early target identification to the optimization of lead compounds.

Researchers can forecast how a tiny molecule will attach to a target protein by using molecular docking. It calculates the interaction profile as well as the binding affinity. By eliminating improbable candidates prior to experimental testing, this method expedites virtual screening. QSAR modeling functions in a different way. It looks for connections between chemical structures and biological activity using statistical and machine learning techniques. QSAR models can forecast the characteristics of novel compounds by being trained on past data. These consist of pharmacokinetics, toxicity, and drug-likeness (Cherkasov et al. 2014).

Pharmacophore modelling indicates the essential characteristics that give a drug its biological activity. These characteristics include aromatic rings, hydrophobic cores, and donors and acceptors of hydrogen bonds. These models help in the creation of compound libraries that correspond to the intended pharmacophore (Lounkine et al. 2012, Schneider et al. 2020).

2. Fundamentals of AI and Deep Learning

2.1 Deep Learning: Definitions and Neural Network Basics

Bioinformatics, systems biology, and cheminformatics have all benefited from these techniques. Deep learning makes use of deep architectures, which are neural networks with numerous hidden layers. Through repeated forward and backward propagation, the model gradually learns significant patterns, increases accuracy, and decreases errors. Deeper networks are able to extract more intricate and abstract properties from unprocessed input data. So, deep learning works particularly well for drug discovery tasks requiring images, sequences, and chemical structures.

2.2 Key Architectures Used in Drug Discovery

Drug discovery relies heavily on a number of Deep Learning (DL) architectures:

Convolutional Neural Networks (CNNs): They are best suited for data that resembles images. 3D molecule structures and their interactions are analyzed using them.

Recurrent Neural Networks (RNNs): They are made to process sequence data. They are used with SMILES string representations of molecules and DNA/RNA sequences.

Generative Adversarial Networks (GANs): Comprising a generator and a discriminator, GANs are employed to create novel drug-like compounds through molecular generation.

Graph Neural Networks (GNNs): GNNs use graphs to represent molecules. They enhance the prediction of molecular properties by capturing spatial and relational information between atoms (Wu et al. 2021).

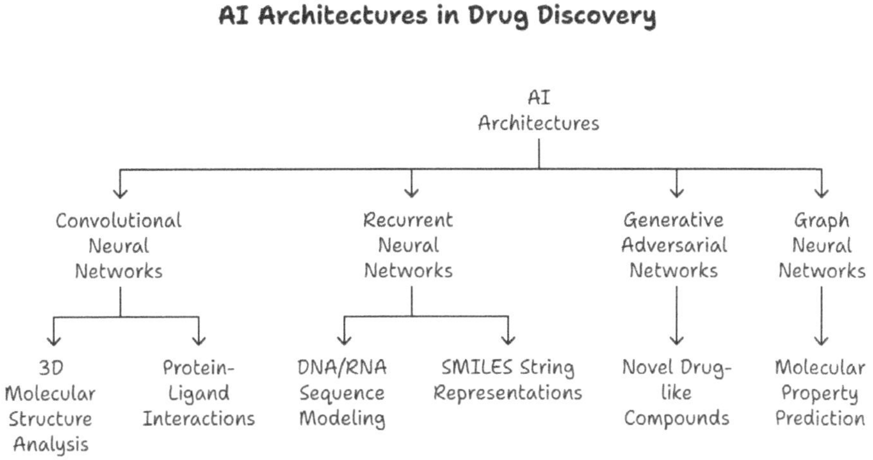

Figure 2: AI architectures in drug discovery.

2.3 Comparison of Traditional ML vs. DL Approaches

Traditional ML methods like decision trees, support vector machines (SVMs), and random forests require domain-specific feature extraction. DL overcomes this by learning hierarchical features directly from raw data, which is especially useful in analyzing biomedical images, molecular graphs, and genomic sequences.

2.4 Tools, Libraries, and Platforms

A range of open-source tools and libraries support AI and DL in drug discovery:

1. **TensorFlow & PyTorch:** Widely used DL frameworks for building and training neural networks.

3.2.4 AI in Drug Repurposing and Repositioning

Drug repurposing is finding new therapeutic uses for existing drugs. It is another domain where AI excels. AI can discover surprising connections between well-known medications and new disease targets by analysing biological data, illness gene expression profiles, and drug target interaction networks using sophisticated computational approaches.

AI is capable of analyzing enormous real-world datasets, such as adverse event reports, EHRs, and clinical trial findings, to find patterns that indicate beneficial off-target effects. Phenotype-based repurposing, in which medications with comparable adverse effects or outcomes are connected to novel possible applications, is also supported.

Especially useful for uncommon diseases or pandemics, repurposed medications can reach patients more quickly because they have previously passed important safety testing. Both development costs and timelines are significantly decreased with this method.

4. Deep Learning Applications in Drug Discovery

4.1 Virtual Screening

Virtual screening is a computational technique used to identify potential hit compounds from large chemical libraries. Deep learning has dramatically increased its speed and accuracy. In deep docking, deep neural networks are used to predict binding affinities based on molecular structures and docking poses, which helps reduce the computational load associated with traditional molecular docking simulations. Ligand-based virtual screening also benefits from deep learning, with models trained to recognize molecular patterns associated with known actives. This allows researchers to rapidly prioritize compounds for experimental validation, reducing costs and improving hit discovery rates. These techniques are especially valuable in large-scale pharmaceutical screening campaigns, where millions of compounds must be evaluated efficiently.

4.2 De Novo Design

In *de novo* drug design, deep learning offers the ability to generate entirely new molecules with desired properties using reinforcement learning (RL) frameworks. Using this technique, a model is trained to create compounds with high reward function scores.

The benefits may include low toxicity, solubility, binding affinity, and ease of production. By making minor adjustments to molecules and using predictive feedback to learn, the RL agent investigates chemical space. This makes it possible to generate therapeutic candidates with specific goals in mind, independent of pre-existing chemical libraries. In cases where rapid and customized chemical discovery is essential—such as for new targets, rare diseases, and rapidly spreading pathogens—reinforcement learning-based design is particularly helpful.

5. Case Studies and Real-World Implementations

5.1 Case Study 1: Benevolent AI – Drug Repurposing for COVID-19

Background and Motivation

Therapeutics were desperately needed in light of the global COVID-19 pandemic. Shorter development times made drug repurposing a feasible approach. BenevolentAI quickly identified approved medications that could fight SARS-CoV-2 by using their AI technology.

Technical Methodology

Using natural language processing (NLP) and graph neural networks (GNNs), benevolent AI explored molecular mechanisms, pathways, and biological targets related to viral infection and immune response.

AI modeled interaction networks using biomedical literature and omics data.
The knowledge graph connected drugs, genes, proteins, and disease mechanisms.

Data and Resources

The system made use of unstructured biomedical text sources as well as structured data from clinicaltrials.gov, EMBASE, and PubMed. Continuous learning pipelines made it possible to integrate real-time COVID-19 literature updates.

Validation and Outcome

Atomwise screened compounds against the polymerase of the Ebola virus in a collaborative research with IBM and the University of Toronto. Several strong inhibitors were suggested and created in less than seven days. More than 600 business and academic partnerships. Effective hit detection in uncommon diseases, parasite infections, and oncology.

Quantitative Outcome

The drug was discovered less than two weeks after the pandemic started. A decrease in ICU hospitalizations and mortality in moderate-to-severe COVID-19 cases. Referenced more than 50 global COVID-19 treatment guidelines.

5.2 Case Study 2: Atom wise – Deep Learning for Virtual Screening

Background and Motivation

Conventional high-throughput screening (HTS) techniques necessitate physical compound libraries and are resource-intensive. Atomwise aimed to find hits more quickly and affordably by using DL-driven virtual screening in favor of wet-lab screening.

Technical Methodology

To predict binding affinity and docking postures, Atomwise's AtomNet employs 3D convolutional neural networks that have been trained on protein-ligand binding data. It takes data from protein crystallography and molecular docking simulations as input. A platform based on bespoke GPU-accelerated algorithms, PyTorch, and CUDA. Able to analyze millions of molecules daily.

Data and Resources

AtomNet was trained on PDBbind, ChEMBL, and in-house compound libraries. Target structures came from PDB and proprietary pharma collaborators.

Validation and Real-World Applications

Atomwise screened compounds against the polymerase of the Ebola virus in a collaborative research with IBM and the University of Toronto. Several strong inhibitors were suggested and created in less than seven days. More than 600 commercial and academic partnerships. Effective hit detection in uncommon diseases, parasitic infections, and oncology.

Quantitative Outcome

- 10 million compounds screened virtually within hours.
- 80% reduction in early-phase screening costs.
- Average hit identification rate: 10x improvement over traditional HTS.

6. Challenges and Limitations

Even though artificial intelligence (AI) and deep learning (DL) have demonstrated encouraging outcomes at different phases of drug discovery—from target identification to molecular generation and repurposing—a number of significant obstacles need to be overcome before these technologies can be completely incorporated into standard pharmaceutical workflows. These difficulties are complex and include data restrictions, ethical concerns, regulatory bottlenecks, practical implementation obstacles, and technological limitations.

6.1 Data Quality, Availability, and Heterogeneity

Large-scale, high-quality datasets are essential for the proper operation of AI models, particularly deep learning architectures. Sadly, the biomedical field is rife with problems pertaining to:

1. Inconsistent Data Formats: Data pertaining to drugs, including omics datasets, chemical structures, biological assays, and clinical trial outcomes, frequently exist in several formats. Harmonization and integration are still technically difficult.
2. Lack of Data in New Domains: There is frequently a severe lack of annotated data for first-in-class targets, rare diseases, or emerging pathogens. This restricts generalization and the use of supervised learning techniques.

3. Biased and Unbalanced Datasets: A number of pharmacological classes, targets, or molecular scaffolds are overrepresented in many of the datasets that are currently available. Predictions may be skewed as a result of AI models trained on such biased data performing badly on underrepresented classes.
4. Proprietary Data Lock-in: Open innovation is hampered by the silos that pharmaceutical companies have created around a vast amount of high-quality data that are not available to the general public or scholarly academics.

6.2 Regulatory Uncertainty and Ethical Concerns

The pharmaceutical industry is already heavily regulated, but AI adds another level of complication. A number of problems emerge:

1. Undefined Regulatory Guidelines: At this time, there is no universal agreement on the best way to evaluate or audit AI models in the context of drug discovery. Preclinical modeling or AI-based biomarker guidelines are still being developed.
2. Data Security and Privacy: AI models that employ patient-derived data (such as gene expression or clinical outcomes) run the danger of breaking data protection regulations like GDPR or HIPAA, particularly if the data is used without the express consent of the patient.
3. Equity and Bias: If AI models are trained on non-representative datasets, they may reflect and potentially exacerbate health inequities. Predictive toxicity models, for instance, that were developed using data from Western populations may not work well for people in Asia or Africa. As a result, various demographic groups may have disparate healthcare outcomes and inaccurate results.

Table 2: Summary Table of Key Limitations.

Category	Specific Challenge	Example Scenario	Proposed Solution
Data	Inconsistent, biased, or scarce data	Rare disease compounds missing in training datasets	FAIR data principles, transfer learning
Interpretability	Black-box model outputs lacking scientific explanation	No rationale for predicted kinase selectivity	Explainable AI (XAI), attention heatmaps
Regulatory	No standardized approval path for AI-derived molecules	FDA delays IND for unexplained AI-suggested lead	Joint regulatory-AI taskforces
Overfitting	Models fail outside training domain	GPCR model misclassifies antibiotic binding	External validation, ensemble models
Experimental Gap	Limited feedback from wet-lab to AI model	Generated compound can't be synthesized	AI-experiment loop platforms, robotic labs
Cost & Skills	Expensive infrastructure and limited interdisciplinary talent	Biotech unable to afford dedicated DL infrastructure	Open-source tools, shared compute resources

Paul, S. M., Mytelka, D. S., Dunwiddie, C. T., Persinger, C. C., Munos, B. H., Lindborg, S. R. et al. (2010). How to improve R&D productivity: The pharmaceutical industry's grand challenge. *Nature Reviews Drug Discovery*, 9(3): 203–214. https://doi.org/10.1038/nrd3078

Schneider, G., and Fechner, U. (2005). Computer-based de novo design of drug-like molecules. *Nature Reviews Drug Discovery*, 4(8): 649–663. https://doi.org/10.1038/nrd1799

Wu, Z., Pan, S., Chen, F., Long, G., Zhang, C., and Philip, S. Y. (2021). A comprehensive survey on graph neural networks. *IEEE Transactions on Neural Networks and Learning Systems*, 32(1): 4–24. https://doi.org/10.1109/TNNLS.2020.2978386

Zhou, Y., Wang, F., Tang, J., and Nussinov, R. (2020). Artificial intelligence in COVID-19 drug repurposing. *The Lancet Digital Health*, 2(12): e667–e676. https://doi.org/10.1016/S2589-7500(20)30200-8

Kadurin, A., Nikolenko, S., Khrabrov, K., Aliper, A., and Zhavoronkov, A. (2017). druGAN: An advanced generative adversarial autoencoder model for de novo generation of new molecules with desired molecular properties *in silico*. *Molecular Pharmaceutics*, 14(9):3098–3104. https://doi.org/10.1021/acs.molpharmaceut.7b00346

Topol, E. J. (2019). High-performance medicine: The convergence of human and artificial intelligence. *Nature Medicine*, 25(1): 44–56. https://doi.org/10.1038/s41591-018-0300-7

Holzinger, A., Carrington, A., and Müller, H. (2022). Measuring the quality of explainability in AI for medicine. *Nature Machine Intelligence*, 4(1): 2–9. https://doi.org/10.1038/s42256-021-00410-x

Rieke, N., Hancox, J., Li, W., Milletari, F., Roth, H. R., Albarqouni, S. et al. (2020). The future of digital health with federated learning. *NPJ Digital Medicine*, 3(1): 1–7. https://doi.org/10.1038/s41746-020-00323-1

Macarron, R., Banks, M. N., Bojanic, D., Burns, D. J., Cirovic, D. A., Garyantes, T.I. (2011). Impact of high-throughput screening in biomedical research. *Nature Reviews Drug Discovery*, 10(3): 188–195. https://doi.org/10.1038/nrd3368

Section 4
Challenges and Opportunities

Chapter 13

Towards Smarter Healthcare

Trends in Quantum Computing, Data Fusion, and Edge AI

Diana George,[1,*] *George Fernandez I.*[1] and *H. Shaheen*[2]

1. Introduction

Healthcare delivery has never been a static entity because every day new innovations are introduced that drastically reshape existing clinical practice. In recent years, the industry has progressed from early digital tools like teleconsultation platforms and electronic health records to a point where intelligent systems actively support diagnosis, continuous monitoring, and personalized treatment planning. Among the several advanced innovations, edge-enabled intelligence, multidimensional information merging, and quantum artificial intelligence (QAI) are the most transformative and promising. By leveraging these new technologies, it is possible to manage the enormous amount of today's complex medical data while providing more individualized, accurate, and timely care suggestions to patients.

The amount of data streams generated by smart wearables, sophisticated imaging, and genomic sequencing is enormous, and existing traditional systems do not have the capability to handle it effectively. Clinicians are also in a dilemma about making quick decisions without compromising patient confidentiality. Although cloud-based AI solutions have shown effective performance in such scenarios, they often face latency issues and raise concerns about the transmission of sensitive data. These challenges have forced the adoption of practical alternatives that implement computational methods beyond the constraints of traditional approaches

[1] Dept of Computer Science and Engineering, Dayananda Sagar University, Harohalli, Banglore, Karnataka.
Email: george.contact@gmail.com, https://orcid.org/0000-0001-6494-9029
[2] Dept of Computer science and Engineering University of West London, Rak Branch Campus, UAE.
Email: H.shaheen@uwl.ac.ae, https://orcid.org/0000-0003-3544-5424
* Corresponding author: dianageorge1994@gmail.com, https://orcid.org/0009-0004-9121-4967

DOI: 10.1201/9781003783855-17

However, these systems were limited to administrative tasks and basic patient management.

By the 2010s, a more connected health ecosystem emerged through cloud computing, mobile health applications, and wearable devices. These technologies provided basic analytics and patient self-monitoring support but lacked integrated intelligence or the ability to fuse multiple data modalities.

The 2020s ushered in a new paradigm by incorporating intelligent, predictive, and real-time healthcare systems driven by Quantum AI (QAI), Edge AI, and multimodal data fusion. These systems enable advanced care by processing and synthesizing heterogeneous data streams from genomics, imaging, clinical records, and biosensors in decentralized and privacy-conscious environments.

Above mentioned Evolutionary progression is summarized in the below given Table 1.

Table 1: Evolution of Healthcare Technologies.

Era	Technology Milestone	Key Characteristics
2000s	EHR, Telemedicine	Basic digitization, electronic records, remote consults
2010s	Cloud AI, mHealth	Mobile apps, wearable sensors, cloud-hosted analytics
2020s	QAI, Edge AI, Data Fusion	Real-time, personalized, privacy-aware intelligence

2.3 Core Features of Smart Healthcare Systems

By enabling clinical intelligence across time and space, smart healthcare environments are gaining popularity.

1. Predictive analytics: Anticipates health issues before they escalate.
2. Preventive care: Uses real-time alerts and early interventions.
3. Personalized treatment: Adjusts plans based on multimodal insights.
4. Participatory health: Engages patients through feedback and monitoring tools.

Hence, using the AI models provide the above-mentioned advantages, which enhance decision-making and enable better treatment options.

2.4 Technical Gaps and Transition Challenges

This section clearly mentions the challenges and existing gaps in the advancement of Quantum AI-based medical care industry.

1. Data Fragmentation: Patient data exists across several sources and is sometimes difficult to connect altogether for interoperability.
2. High Latency: Cloud-based centralized models often struggle to offer real-time insights when needed due to processing delays.
3. Privacy and Compliance: Exposure of highly sensitive personal data to unsecured network environments is unsafe, and systems must adhere to legal frameworks (HIPAA, GDPR) and ethical norms.

4. Skill Deficit: Many healthcare organizations and individuals lack knowledge about these infrastructures, as the domain is interdisciplinary and requires workforce expertise in deploying edge, AI, or quantum solutions.

To overcome these aforementioned challenges, healthcare systems incorporate multimodality fusion. A simple mathematical formula representing this fusion is shown in equation (1)

$$\hat{y} = f(concat(x_{ehr}, x_{imaging}, x_{sensor}, x_{genomics})) \quad (1)$$

Where, $\hat{y}$ is the Predicted output, $x_{ehr}, x_{imaging}, x_{sensor}, x_{genomics}$ are categorized as feature vectors from each modality, and means the Concatenation or fusion of modalities before input to make the predictive model f. For the smart diagnostics and personalized care, this system acts as the key to multimodal integration.

2.5 Role of Data in Driving Smart Healthcare

In this contemporary era, data acts as the lifeblood of intelligent systems because from wearable sensors to next-generation sequencing, the variety and granularity of data sources need to explored.

To summarize, the evolution of the healthcare system from the era of digitization to intelligence has been enhanced due to the explosion of medical data and the need for time-bounded processing of such data to make quick decisions. In the bygone days, systems focused on record-keeping and remote access, but current smart healthcare platforms are capable of predicting, personalizing, and preventing disease through seamless integration of diverse data modalities mentioned in Table 2. In spite of significant progress, challenges still remain, such as fragmented data silos, latency in centralized AI systems, and regulatory constraints, all of which limit scalability. Quantum-powered computation, multimodal data fusion, and intelligent edge deployment act as promising solutions for the future, and these technologies will together form the backbone of a truly responsive, adaptive, and future-ready healthcare ecosystem.

Table 2: Healthcare Data Modalities and Their Roles in Smart Systems.

Data Type	Example Source	Format	Use in Smart Health
EHR	Hospital Systems	Structured tables	Medical history, medications
Imaging	MRI, CT, X-ray	Pixel matrices	Diagnosis (e.g., tumor, stroke)
Wearables	Fitness Trackers	Time-series	Heart rate, sleep cycles, oxygen levels
Genomics	DNA Sequencing Labs	Symbolic strings	Personalized treatment, risk prediction

3. Quantum Artificial Intelligence in Healthcare

3.1 The Rise of Quantum AI in Medicine

To overcome the current data-handling inefficiencies of traditional systems, more promising Quantum Artificial Intelligence (QAI)—the combination of quantum computing and artificial intelligence—is required for computational diagnostics and medical accuracy. By using quantum concepts like superposition, entanglement, and quantum interference to assess enormous solution spaces, QAI performs its tasks very quickly, making it suited to the modern world of massive, complex, and multidimensional health data. This emerging technology also facilitates quicker diagnosis, accurate therapy prediction, and deeper understanding of clinical imaging and genetic data in medical settings.

3.2 Core Principles of Quantum Computing

Quantum computing differs from the traditional binary zero-and-one concept by enabling multiple simultaneous states. A quantum bit, or qubit, can exist in multiple states at a time, enabling massive parallelism. Entangled qubits maintain correlations regardless of distance, while interference refines results.

Superposition state of a qubit can be defined as, $|\varphi\rangle = \alpha|0\rangle + \beta|1\rangle$ (2)

Where, $|\alpha|^2 + |\beta|^2 = 1$ (3)

These operations are implemented through quantum gates like Hadamard, Pauli-X, and CNOT, forming quantum circuits analogous to neural networks.

3.3 Key Quantum Machine Learning Techniques

This section focuses on the QAI workflow as shown in Figure 1 and how it extends traditional ML algorithms, its advantages and the application mentioned in the Table 3:

1. Quantum Support Vector Machines (QSVM): Replace classical kernels with quantum ones for better separation in feature space.

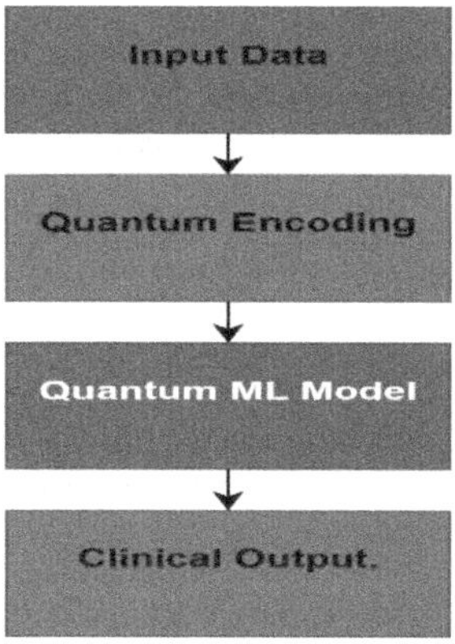

Figure 1: Quantum AI Workflow in Healthcare.

Table 3: Application of Quantum Computing in Healthcare Industry.

Application Area	Quantum Advantage	Healthcare Impact
Drug Discovery	Quantum molecular simulation	Faster screening of therapeutic compounds
Genomic Analysis	Clustering high-dimensional genomic sequences	Early disease susceptibility prediction
Radiological Imaging	Enhanced reconstruction and segmentation	More accurate and faster image analysis
Optimization in Diagnosis	Real-time treatment planning using VQC	Timely and personalized clinical decisions

2. Variational Quantum Circuits (VQC): Hybrid models trained using classical optimization but executed on quantum hardware.
3. Quantum Principal Component Analysis (qPCA): Speeds up dimensionality reduction of large datasets.
4. Quantum Boltzmann Machines: Quantum analogues of deep learning models for generative tasks.

3.4 Medical Applications of Quantum AI

QAI is not just theoretical—it is being explored in diverse clinical applications:

3.5 Hybrid Quantum-Classical Architectures

Current quantum hardware (known as **NISQ** devices) has limitations in qubit count and error tolerance. Therefore, most QAI models are hybrid systems, combining classical and quantum computation because classical modules perform data pre-processing and initial transformation and quantum modules execute learning or optimization tasks. This collaborative model allows developers to harness quantum advantages without sacrificing practical deployment. The processing time can be calculated using the following equation (4):

$$T_{Total} = T_{Pre}^{(classical)} + T_{Encode}^{(quantum)} + T_{Infer}^{(quantum)} \tag{4}$$

Frameworks like Qiskit, PennyLane, and TensorFlow Quantum simplify these integrations and allow simulation even without access to quantum hardware.

3.6 Quantum Data Encoding & Readout

Medical data must be encoded into quantum states using methods such as:

1. **Basis Encoding:** Maps binary features (e.g., symptoms).
2. **Amplitude Encoding**: Embeds continuous data (e.g., lab values).
3. **Angle Encoding:** Rotates qubit gates based on input features (e.g., wearable signal).

2. HER's historical data on medication and cholesterol
3. Cardiac MRI image features

This multimodal integration allows for more accurate early detection of arrhythmia than ECG-only systems

B. In oncology, advanced detection combining histopathological images, genomic variants, and radiological scans helps assess tumor grade, metastasis potential, and predict treatment response.

4.6 Handling Missing or Noisy Modalities

Multimodal systems face difficulties when one or more input streams are unavailable or corrupted like:

- Imputation techniques can fill missing data (e.g. KNN, GAN-based generation).
- Cross-modal inference allows estimating one modality from another (e.g. generating missing CT from MRI).
- Hybrid fusion models with dropout-aware training can gracefully degrade performance.

4.7 Explainability in Multimodal Models

In order to validate the medical outcomes effectively, clinicians require trustworth interpretable models:

- Saliency maps can highlight issue causing part of image regions.
- Attention weights across modalities reveal how much each modality contributes to the effectiveness of given predictions.
- Shapley values or feature attribution methods provide transparency in EHR fusion.

4.8 Performance Metrics for Fusion Systems

In order to integrate various medical data streams like EHRs, imaging, genomics, and sensor data into integrated models for more precise, context-aware, and time-limited decision-making, heterogeneous data consolidation serves as an essential part. Attention or cross-modal interaction inside deep neural network frameworks are the most sophisticated hybrid fusion techniques. Early fusion strategies enable simplicity by combining features prior to modeling, whereas late fusion aggregates results from several models. Diseases like cardiology, oncology, and intensive care unit monitoring have utilized the advantages of this multimodality fusion and its advanced individualistic supportive therapy sessions. Table 6 mentioned some of the common performance metrics and its purpose for the easy reference.

Table 6: Performance metrics for the QAI.

Metric	**Purpose**
AUC-ROC	Discrimination power
F1-Score	Balance of precision and recall
Modality Robustness	Performance under missing/noisy data
Interpretability Score	Qualitative rating from clinicians

5. Edge AI: Healthcare at the Point-Of-Care

Edge AI acts as a very crucial disruptive tool in terms of continuous or enormous data collection and its processing stage, with the decentralization of healthcare. It empowers end devices such as wearables, monitors, and mobile diagnostics to perform locally for improved inference and data protection compared to the traditional round-tripping of cloud systems. This architecture significantly lowers bandwidth requirements, reduces latency, and enhances autonomy, and these features make this QAI system ideal for time-sensitive and resource-constrained environments and it is shown in Figure 2.

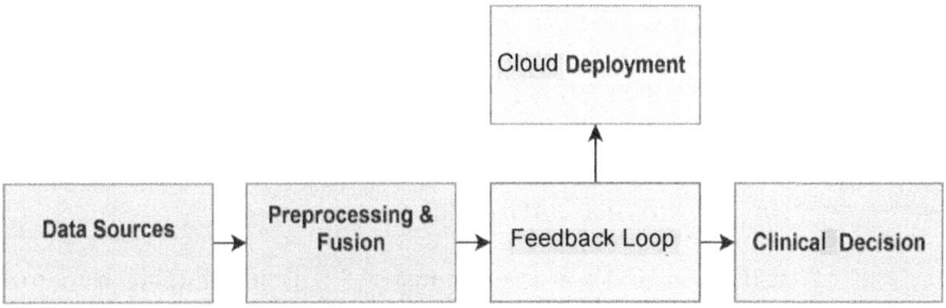

Figure 2: Edge AI Deployment Framework.

5.1 Fundamentals of Edge AI in Healthcare

Edge AI refers to deploying machine learning models on local hardware at the edge instead of relying solely on remote cloud servers. Edge nodes can be medical IoT devices, smartphones, embedded boards, or specialized low-power accelerators.

Key characteristics of Edge AI:

1. On-device inference: AI model runs locally without round-trips to cloud
2. Low latency: Immediate response time, crucial for alerts and emergency care
3. Data minimization: Sensitive patient data is kept on-device, reducing risk
4. Energy efficiency: Optimized for constrained environments (battery-operated, solar-powered)

5.2 Enabling Technologies: Edge Hardware Platforms

Several hardware platforms have been designed specifically for healthcare Edge AI deployment with the specification mentioned in Table 7:

These devices support on-device inference for tasks such as anomaly detection in ECG, oxygen level monitoring, seizure prediction, and remote diagnostics in telehealth.

5.3 Federated Learning for Privacy-Preserving Training

One of the critical breakthroughs enabling collaborative AI on edge devices is federated learning (FL). In FL, multiple edge devices train a shared model locally on their private datasets. Instead of transferring raw data to a central server, only model updates (gradients or weights) are aggregated and synchronized.

$$w_{t+1} = w_t + \eta \sum_{i=1}^{N} \frac{n_i}{n} \nabla n_i \tag{6}$$

Where w_t is the global model at time t, ∇w_i is the gradient from device i, n_i is the data sample count on device i, and $n = \sum_i n_i$, and η is the learning rate.

This ensures greater privacy, the model adapts to regional or patient-specific traits, supports thousands of distributed nodes.

Table 7: Hardware specifications.

Platform	Hardware Specs	Use Case	Power Usage
Raspberry Pi 4	ARM Cortex-A72, 4GB RAM	Local monitoring, portable diagnosis	~ 6W
Coral Edge TPU	TensorFlow Lite accelerator	Fast medical image inference	~ 2W
NVIDIA Jetson Nano	GPU-accelerated edge AI	High-resolution real-time imaging	~ 5–10W
Arduino Portenta	Ultra-low-power MCU + AI chip	Biosignal tracking from wearables	< 1W

5.4 Applications of Edge AI in Healthcare

Edge AI is being used in various real-world scenarios and some of the cases are mentioned in Table 8 for better understanding.

These applications demonstrate how edge intelligence improves diagnostic continuity and minimizes dependence on continuous connectivity.

5.5 System Integration and Edge-Cloud Hybrid Models

Edge AI does not aim to replace cloud computing but rather complement it. In hybrid architectures, the edge performs initial analysis and alerts (low-latency), and the cloud aggregates global trends, retrains models, and supports long-term analytics.

Table 8: Real life applications of Edge-AI.

Use Case	Description
Smartwatches for ECG	Real-time arrhythmia detection and alerts on-device (e.g., Apple Watch, Fitbit)
ICU Bedside Monitors	Local processing of vitals for anomaly detection and escalation
Rural Clinics	Mobile diagnostic tools using Raspberry Pi and low-power models
Post-operative Monitoring	Home-based health monitoring with wearables

5.6 Benefits and Limitations

(1) Benefits:
  a) Real-time decision-making
  b) Offline operation in remote areas
  c) High data privacy and patient control
  d) Scalable and adaptive models

(2) Limitations:
  a) Model compression and accuracy trade-offs
  b) Hardware constraints (limited memory and compute)
  c) Complex deployment pipelines (cross-platform dependencies)
  d) Energy consumption in continuous monitoring setups

5.7 Benchmarking Edge AI Performance in Healthcare

The inference accuracy of AI models always depends on how fast and precisely they can process data under resource constraints while implementing them on edge devices. This is important in the healthcare industry because the decision-making latency might be very crucial. The inference performance of prominent AI models installed on well-known edge hardware platforms is summarized Table 9 and given below.

It is clear from the above given data findings that well-optimized deep learning models for diagnostic inference on edge platforms are feasible. For example, real-time seizure detection and prediction using CNN algorithms with the support of

Table 9: Benchmark Performance of Healthcare AI Models on Edge Devices.

Model	Device	Inference Latency (ms)	Accuracy (%)	Use Case
MobileNetV2 (TFLite)	Coral Edge TPU	4.2	88.7	Skin lesion classification
ResNet18	Jetson Nano	19.8	92.3	Chest X-ray diagnosis
CNN-1D (EEG)	Raspberry Pi 4	9.5	91.1	Seizure prediction

5. Feedback and Clinical Decision Layer

This last layer transforms AI outputs into clinically actionable insights for informed decisions:

a) Dashboards: Risk scores, image overlays, time-series alerts
b) Integration: Links with hospital EMR systems
c) Feedback Loop: Adjustments that help retrain the AI pipeline periodically

Apart from this, interpretability tools like on-device saliency maps or attention heatmaps support medical practitioners in understanding the basis of AI recommendations, improving trust and transparency.

7. Applications and Case Studies

The delivery of medical services has been transformed since the incorporation and introduction of Quantum Artificial Intelligence (QAI), multimodal data fusion, and Edge AI into the medical sector. Four important areas are covered here for a better understanding of the usage of these technologies. These include the management of chronic diseases, detecting cancer, responding to pandemics, and rural healthcare.

7.1 Pandemic Response: COVID-19 and Beyond

Systemic flaws in healthcare response systems were made clear by the COVID-19 pandemic because that time urged for rapid diagnosis, decentralized alert systems, and real-time observation. Some similar applications and the technology used during the implementation of such application are mentioned in Table 10.

Table 10: Summary of Applications and Technologies.

Use Case	Technology Used	Benefit Achieved
COVID-19 Detection	Edge AI + FL + CNN	Instant triage, privacy-preserving models
Cancer Screening	QAI + Image Fusion	Higher sensitivity in early detection
Epilepsy Monitoring	1D-CNN on Edge	Real-time alerts with <10 ms latency
Diabetes Optimization	QAI + RL + EHR Fusion	Personalized insulin dosing
Rural Health Screening	Edge AI + Offline Fusion	Diagnosis without cloud or specialists

7.1.1 Application of Smart Technologies: diagnostic synergy, robustness, contextual depth personsalization and real-time decisions

a) Edge AI-powered thermal imaging systems were deployed at airports and hospitals to detect febrile individuals in real time.
b) Federated Learning models were used across hospitals to train robust COVID-19 classifiers without sharing patient data.
c) QAI models (simulated using hybrid quantum circuits) showed promise in optimizing contact tracing networks and drug repurposing.

7.2 Cancer Detection and Image-Based Diagnostics

Radiological scans, pathology slides, and the integration of the genomic profile are the main tools used for cancer diagnosis, as it is a strong contender for multimodal AI systems because of its complexity.

Smart Fusion Use Case:

a) PET-CT Fusion: Combines metabolic and anatomical views for better tumor localization.
b) QAI Application: Quantum-enhanced SVMs were tested on high-dimensional gene expression data to classify cancer subtypes with improved feature efficiency.
c) Edge Use Case: Skin lesion classifiers (e.g., Mobile Net on Coral TPU) enabled offline triage in dermatology screening programs.

7.3 Chronic Disease Management and Home Monitoring

Continuous monitoring and adaptive decision-making with the help of Edge AI are extremely beneficial for chronic illnesses, especially diabetes, epilepsy, and hypertension.

7.3.1 Use Cases:

a) EEG-based seizure prediction using 1D-CNNs deployed on Raspberry Pi.
b) Federated learning models trained on wearable ECG data to flag early signs of arrhythmia.
c) Personalized dosage optimization via quantum-enhanced reinforcement learning for Type 1 diabetes patients.

By completing the cycle of data gathering, real-time analysis, and therapeutic action, these technologies enable proactive home care where required.

7.4 Rural and Resource-Constrained Settings

Closing the healthcare gap between rural poor and urban areas is believed to be the most important use of smart healthcare. Even though there is a lack of permanent specialists in rural areas, they might be outfitted with reasonably priced AI edge instruments.

Technological Enablers:

a) Offline inference on mobile devices for tuberculosis detection from chest X-rays.
b) Lightweight fusion engines that work without internet, processing imaging and basic vitals.
c) QAI use (cloud-assisted): AI-enhanced teleradiology systems with quantum-optimized image prioritization queues for remote diagnosis centres.

7.5 Insights and Impact

These case studies demonstrated that no technology functions in isolation because it has been seen that the most significant deployments occur when Edge AI manages real-time limitations, QAI addresses deep optimization or complexity, and data fusion guarantees contextual richness efficiently. These tools work together to change healthcare access and precision techniques, democratizing disease screening in the wake of pandemic preparedness. This will efficiently follow the use cases that will no longer be experimental but will define normal medical protocols in the upcoming years when hardware advances and quantum systems scale, primarily in the areas of telemedicine, emergency medicine, and preventative care.

8. Ethical, Legal, and Social Implications (ELSI)

Along with the availability of several scientific challenges, the integration of the most promising cutting-edge medical technologies like Edge AI, multimodal fusion, and quantum AI will attract several other critical challenges in regard to significant ethical, legal, as well as social concerns. Even though these emerging technologies have huge potential, their use in sensitive medical settings necessitates data governance, justice, transparency, and societal equity as mentioned in the ethical traingle shown in Figure 4.

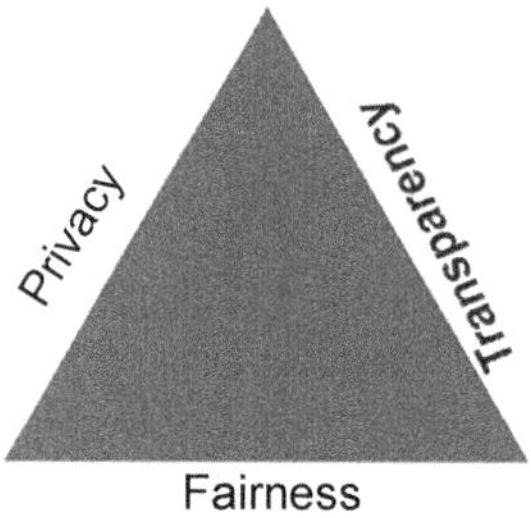

Figure 4: Ethical AI Triangle.

8.1 Data Privacy and Security

The privacy and safety of the patient must be the first priority whenever AI models access genomic data, electronic health records (EHRs), as well as real-time physiological signals because they deal with personal information. In this regard, Edge AI reduces the need to send raw patient data to centralized servers by keeping computation local. Additionally, Federated Learning presents a feasible method for AI training by protecting privacy by sharing only model weights instead of sending raw patient data.

However, some risks still remain, like membership inference attacks and any kind of data leakage through model inversion. In this scenario, it is better to transfer the data in encrypted format; secure multiparty computation (SMC) and differential privacy also need to be incorporated into system designs. Also, regulatory frameworks and conducting all kinds of experiments in compliance with laws like

HIPAA (USA), GDPR (EU), and DISHA (India) should be non-negotiable in any deployment involving personal health data security.

8.2 Algorithmic Bias and Fairness

Sometimes trained AI models on imbalanced datasets may have the tendency to perpetuate or even amplify biases in clinical decision-making. Mitigating these bias issues requires:

1. Diverse, representative datasets across age, gender, ethnicity, geography.
2. Explainability frameworks (e.g., SHAP, attention heatmaps).
3. Continuous monitoring of model outputs for fairness indicators.

8.3 Social Equity and Access

The aim of all emerging technologies should be reducing the digital divide. Therefore, it ensure that people are not left behind due to their low income or limited access to techology:

1. Lack of edge-compatible devices
2. Limited internet bandwidth for federated updates
3. Language or other kind of literacy barriers in app interfaces

8.4 Looking Ahead

Ethical AI is not a post-processing step but must be considered from the architecture viewpoint to the training data and deployment plan. So, ethics related to these implementations also need to shift from scholarly discussion platforms to an engineering table with the proliferation of QAI and Edge AI. However, interdisciplinary cooperation between technologists, physicians, ethicists, and legislators is required to create reliable, equitable, and inclusive healthcare platforms.

9. Challenges and Research Issues

In spite of the benefits and potential of combining Quantum AI (QAI), Edge AI, and multimodal data in healthcare, there are major challenges when it comes to practical implementation, including computing, infrastructure, ethics, as well as operations.

9.1 Quantum AI: Hardware and Scalability Barriers

Despite the theoretical benefits of QAI, the application of quantum computing in healthcare is still in its infancy due to implementation bottlenecks like:

1. Limited Qubit Fidelity: Current quantum processors are noisy and prone to decoherence, affecting result reliability.
2. Restricted Qubit Count: Most available quantum systems (e.g., IBM Q, Rigetti) are limited to 50–100 qubits, which are insufficient for large medical datasets.

Akahoshi, Y., Fujisaki, J., Oshima, H., Sato, S., and Fujii, K. (2024). General-purpose quantum circuit generator for partially fault-tolerant quantum computing architecture. *2024 IEEE International Conference on Quantum Computing and Engineering (QCE)*, 420–421. https://doi.org/10.1109/QCE60285.2024.10335

Baltrušaitis, T., Ahuja, C., and Morency, L. P. (2019). Multimodal machine learning: A survey and taxonomy. *IEEE Transactions on Pattern Analysis and Machine Intelligence*, 41(2): 423–443. https://doi.org/10.1109/TPAMI.2018.2798607

Basak, J., Wang, B., and Rajasekaran, S. (2025). Problem specific communications in distributed quantum computing. *2025 International Conference on Quantum Communications, Networking, and Computing (QCNC)*, 406–410. https://doi.org/10.1109/QCNC64685.2025.00069

Biamonte, J., Wittek, P., Pancotti, N., Rebentrost, P., Wiebe, N., and Lloyd, S. (2017). Quantum machine learning. *Nature*, 549: 195–202. https://doi.org/10.1038/nature23474

Brooks, C. (2025). Artificial intelligence and health care. In *Inside cyber: How AI, 5G, IoT, and quantum computing will transform privacy and our security* (pp. 147–150). Wiley. https://doi.org/10.1002/9781394310562.ch18

Hasan, M. M., Rahman, M. M., Ali, M. M., and Machado, P. (2024). QuantoTrace: Quantum error correction as a service for robust quantum computing. *2024 6th International Conference on Electrical Engineering and Information & Communication Technology (ICEEICT)*, 616–621. https://doi.org/10.1109/ICEEICT62016.2024.10534391

Jobin, A., Ienca, M., and Vayena, E. (2019). The global landscape of AI ethics guidelines. *Nature Machine Intelligence*, 1: 389–399. https://doi.org/10.1038/s42256-019-0088-2

Kumar, P. V., Kulkarni, A., Mendhe, D., Keshar, D. K., Tilak Babu, S. B. G., and Rajesh, N. (2024). AI-optimized hardware design for Internet of Things (IoT) devices. *2024 5th International Conference on Recent Trends in Computer Science and Technology (ICRTCST)*, 21–26. https://doi.org/10.1109/ICRTCST61793.2024.10578352

McMahan, H. B., Moore, E., Ramage, D., Hampson, S., and y Arcas, B. A. (2017). Communication-efficient learning of deep networks from decentralized data. *Proceedings of AISTATS*. https://arxiv.org/abs/1602.05629

Mittelstadt, B. D., Allo, P., Taddeo, M., Wachter, S., and Floridi, L. (2016). The ethics of algorithms: Mapping the debate. *Big Data & Society*, 3(2): 1–21. https://doi.org/10.1177/2053951716679679

Mohammed, A. A. A., Obaid, A. M. H., Alsabry, A., F. S. A., and Alhaj, Y. A. (2024). Exploring the impact of quantum internet on telemedicine technology: Enhancing healthcare delivery, speed, and data security. *2024 1st International Conference on Emerging Technologies for Dependable Internet of Things (ICETI)*, 1–9. https://doi.org/10.1109/ICETI63946.2024.10777189

Nguyen, H.-S., and Voznak, M. (2024). A bibliometric analysis of technology in digital health: Exploring health metaverse and visualizing emerging healthcare management trends. *IEEE Access*, 12: 23887–23913. https://doi.org/10.1109/ACCESS.2024.3363165

Rajesh, R., Hemalatha, P., Vijayabharathi, R., Bargunan, S., Ramprasad, S. R., and Raja, J. (2025). Quantum-enhanced deep learning for predictive diagnostics and personalized treatment in smart healthcare systems. *2025 5th International Conference on Pervasive Computing and Social Networking (ICPCSN)*, 794–801. https://doi.org/10.1109/ICPCSN65854.2025.11035338

Roy, Y., Banville, H., Albuquerque, I., Gramfort, A., Falk, T. H., and Faubert, J. (2019). Deep learning-based electroencephalography analysis: A systematic review. *Journal of Neural Engineering*, 16(5): 051001. https://doi.org/10.1088/1741-2552/ab260c

Schuld, M., Sinayskiy, I., and Petruccione, F. (2014). The quest for a quantum neural network. *Quantum Information Processing*. https://arxiv.org/abs/1408.7005

Sutradhar, K., Venkatesh, R., and Venkatesh, P. (2024). A review on smart healthcare employing quantum Internet of Things. *IEEE Engineering Management Review*. https://doi.org/10.1109/EMR.2024.3440835

Ullah, M. A., Awan, A. J., and Svensson, E. (2024). Towards compute capacity maximization in constrained interconnect multi-chip quantum computing. *2024 IEEE International Conference on Quantum Computing and Engineering (QCE)*, 368–369. https://doi.org/10.1109/QCE60285.2024.10309

Zheng, J., Luo, M., Zhong, Y., and Chen, Z. (2024). Generative AI-based short-term electricity load forecasting in 5G mobile edge computing. *2024 3rd International Conference on Computing, Communication, Perception and Quantum Technology (CCPQT)*, 95–100. https://doi.org/10.1109/CCPQT64497.2024.00025

Chapter 14

Applications of AI in Clinical Biochemistry Laboratories

A Review

Ranjay Kumar Choudhary,[1] *Prasad Khodke,*[2] *Renu Chane,*[3] *Saksham Saxena,*[1] *Nuwo David,*[4] *Vivek Kumar Garg,*[5,*] *Shubham Koundal,*[6] *Bhavya Trivedi*[7] and *Pankaj Kumar*[8]

1. Introduction

Recent advances in artificial intelligence (AI) computer programs that simulate the intelligence of humans or cognitive mechanisms have prompted optimism that AI could enhance the outcomes of patients and improve the efficacy of health care

[1] Department of Medical Laboratory Sciences, College of Applied and Health Sciences, A'Sharqiyah University, 400, Ibra, Sultanate of Oman.
Emails: r.choudharymt@gmail.com, https://orcid.org/0000-0003-2801-4079; sakshamsaxena0608@gmail.com, https://orcid.org/0009-0009-8924-8662

[2] Department of Biochemistry, Nootan medical College and Research Center, Sankalch and Patel University, Visnagar, Gujarat, India-384315.
Email: prasadkhodke88@gmail.com, https://orcid.org/0000-0001-6632-0611

[3] Department of Biochemistry, School of Medical Sciences and Research, Sharda University.
Email: renuchane@gmail.com, https://orcid.org/0000-0003-3470-5386

[4] Department of Medical Lab Technology, University Institute of Allied Health Sciences, Chandigarh University, Mohali, Punjab.
Email: nuwodavid81@gmail.com, https://orcid.org/0009-0000-6960-4281

[5] Department of Medical Lab Sciences, USAHS, Rayat-Bahra University, Mohali Punjab, India-140104.

[6] Chandigarh Group of Colleges, Jhanjeri, Mohali, Punjab,140307, Chandigarh School of Business, Department of Medical Laboratory Sciences.
Email: rohitkoundal79@gmail.com, https://orcid.org/0000-0003-0608-3882

[7] School of Applied & Life Sciences, UIT, Uttaranchal University, Dehradun, India-248007.
Email: bhavya.trivedi25@gmail.com, https://orcid.org/0000-0001-8292-8491

[8] Department of Medical Biochemistry, Index Medical College, Hospital and Research Centre, Malwanchal University, Indore, MP, India.
Email: pankajmarshal@gmail.com, https://orcid.org/0009-0007-6234-668X

* Corresponding author: garg.vivek85@gmail.com, https://orcid.org/0000-0002-7906-2153

DOI: 10.1201/9781003783855-18

delivery. AI-powered hospitals and lab facilities are capable of executing a broad spectrum of roles, such as diagnosis, completely or partially on their own (Pashkov et al. 2020). Patients and the wider population could possibly monitor themselves and diagnose themselves for conditions like atrial fibrillation, skin imperfections, and retinal illnesses using wearable devices and applications with AI capabilities. The perspectives and requirements of impacted participants surrounding the implementation of AI in healthcare have not yet been fully known, partly because of the constantly changing characteristics of contemporary AI systems. With the deployment of sophisticated instrumentation and integrated processes that incorporate modern machinery and programming with artificially intelligent capabilities, clinical testing facilities have been growing progressively automated. Computerized machines can now swiftly and efficiently complete obligations that had previously been done by laboratory workers or technologists because of technological advances in automation (Isaacs 2020). The examination volume may be substantially increased through the use of contemporary computerized technologies that coordinate the preservation of samples through assay analysis. The discipline of ML is often categorized broadly into two distinct subcategories: unsupervised and supervised. The most common application of AI in medicine, supervised ML, involves a computer recognizing trends in already labeled data, wherein the desired label is known. By providing the machine's program feedback on its proper response, these markings assist the algorithm in generating better predictions. The use of support vector machines, tree-based designs, logistic and linear regression techniques, as well as additional supervised artificial intelligence models, can all be suitable for a range of healthcare tasks. In this review of the literature, tree-based techniques were frequently employed, and these algorithms have generally performed well in medical applications. A tree of choices composed of a complex sequence of selection points is employed within these models (Yu et al. 2018a). Throughout the model development process, the decision points are deduced from the training data. Unsupervised learning in ML, on the other hand, requires providing algorithms an unlabeled dataset. The framework continues to use the trends or patterns it finds to explain the relationships in the data. Unexpected patterns may be discovered by ML that is unsupervised. The analysis of principal components, k-means clustering, and k-nearest neighbors are a few types of unsupervised AI methods. In conclusion, it's essential to contextualize DL and AI with ML. The broader field of employing machines to carry out human-like operations, such as resolving issues, has been referred to as AI. AI includes machine learning as a specialization. The field of deep learning, on the other hand, is a branch of ML that draws inspiration from neural networks in the brain. Each artificial neural network layer in the algorithms for DL expands upon the one that comes before it to glean ever-more intricate insights from the data that comes in. These approaches' emergent quality explains their increasing attractiveness in the healthcare sector, because they allow the successful completion of difficult assignments like language interpretation and picture recognition (Pashkov et al. 2020). The increasing need for accuracy, effectiveness, and dependability in clinical biochemistry labs is what spurred this investigation. There is growing demand on the world's healthcare system to provide quick, affordable, and high-quality diagnostic services. Conventional manual and semi-automated methods are frequently insufficient for fast and error-free analysis due to the enormous volume of biological data created every day. By automating repetitive procedures, forecasting and confirming results, and assisting physicians in making decisions, artificial intelligence (AI) offers potent answers. Additionally, by integrating multi-modal data, it opens up new avenues for predictive modeling and tailored medication, bringing laboratory practice into line with the way healthcare will be delivered in the future.

2. Methodology

With the goal of discovering possibly significant study publications that investigate the Applications of AI in Clinical Biochemistry Laboratory, a literature assessment was completed using Google Scholar, PubMed, the Scopus search engine, the Digital Library of Cochrane, and other databases. AI, Clinical Biochemistry, Laboratory, and Patient care were the keywords, search phrases, and subject headers used in both the PubMed and Google Scholar search engines. In an effort to reduce finding biases, the gathered list of references was carefully examined for relevant works to identify new references that might be overlooked.

3. Literature Review

Our analysis of ML's present state in laboratory medicine identifies a number of fascinating uses, such as predicting test results, enhancing lab utilization, automating lab procedures, encouraging accurate test interpretation, and, lastly, enhancing laboratory medicine information systems. The best results in these experiments were frequently obtained using neural networks and tree-based learning techniques.

4. Laboratory Optimization and Forecasting of Test Values

Prediction of laboratory findings focused on other clinically available information has become a common theme in the utilization of AI in medical laboratory science, as shown in Figure 1. In accordance with the researchers of these assessments, these models are capable of being utilized to maximize the implementation of tests performed in laboratories and empower clinical decision-support instruments for purchasing clinicians, illustrated in Figure 2. The research conducted by Azarkhish et al., which utilized characteristics from a regular complete blood count to estimate anemia caused by iron deficiency and serum iron levels, constitutes one of the very first instances of this particular type of work (Kocak 2024). The model's 98% AUROC for the categorical classification of anemia caused by iron deficiency was attained. A healthcare decision-support system that can predict laboratory test outcomes based on associated laboratory findings and other clinical data has been established as part of this ongoing effort. As a proof of concept, they demonstrated that an automated learning algorithm could determine with significant accuracy whether or not a serum ferritin level was abnormal (Warraich et al. 2025). They accomplished an AUROC score of 97% by completing the necessary lacking laboratory characteristics utilizing the randomly generated forest imputation method, which at first was subsequently introduced into a model based on logistic regression. On the other of 100%, the algorithm successfully categorized important parameters as legitimate. This implies that the model could substantially reduce the quantity of critical findings that require validation by hand yet still preserve a small proportion of tests that have been improperly validated. In a comparable manner, a researcher designed an array of tree-based methods to automate the procedures of validating test results (Alami et al. 2021). The algorithm they used had a 99 percent sensitivity and a 98% specificity in automatically validating laboratory data. Relative to their existing rule-based verification technique, this could have brought about an 80 percent reduction of laboratory findings requiring verification by hand due to their historical data. In the meantime, fewer specimens have been highlighted for manual inspection utilizing a tree-based neural network model (Jocelyn Chew and Achananuparp 2022, Smith et al. 2020). In comparison to expert-labeled ground truth manually induced urine microscopic results, the model they developed, which utilized characteristics from a 10-point dipstick and urine cytometry data, necessitated a manual evaluation rate of 32 percent, which translated into a sensitivity rating of 92 percent and specificity of 81.5 percent. Reports demonstrated that an automated machine learning system to identify incorrectly identified samples from laboratories could surpass human verification in terms of quality assurance/quality control (Hanna et al. 2025). A network of neural networks that had an AUROC of 98 percent was their top-performing algorithm. Nevertheless, this study's inability to compare its effectiveness to the moment's standard of excellence, rules-based delta checks, is a shortcoming. In the meantime, a neural network machine learning algorithm can moderately accurate (AUROC 91%) determine if a blood sample was clotted or not. The method they used compared the results of models to the fundamental truth of being manually examined for coagulation by laboratory technicians using the specimen's coagulation test findings. Farrell et al. demonstrated that a machine learning method for detecting incorrectly labeled lab samples could outperform manual verification in terms of quality assurance/ quality control. A neural network with an AUROC of 98% was their top-performing algorithm. However, this study's failure to compare its performance to the current gold standard, rule-based delta checks, is a shortcoming. In the meanwhile, Fang et al.'s neural network machine learning system demonstrated a moderate level of accuracy (AUROC 91%) in determining if a blood sample was clotted or not. Their approach compared model outputs to a ground truth of manual examination for clotting by laboratory technicians using the sample's coagulation test findings.

7. Laboratory Information Systems

In an effort to strengthen operations and facilitate clinical investigation, machine learning has additionally been implemented in laboratory information systems. For example, in an effort to promote clinical investigations and seamless integration, two studies in this review suggested employing machine learning approaches for converting laboratory information into standard LOINC codes (Hanna et al. 2025). The industry-standard ontology for conveying information about measurements in medical applications, particularly tests performed in laboratories, is now LOINC codes. One such research paper offers a set of frameworks for translating seven typical laboratory concepts to the health information system of the United States Department of Veteran's Affairs, wherein LOINC mappings are not ideal. With an accuracy of 98 percent, a tree-based approach performed the best, providing an important advantage over the time-consuming task of manually reviewing hundreds of potential conceptual relationships. In the United States Veterans medical records data warehouse, a model using machine learning has been developed to assign lacking LOINC identifiers and enhance the level of accuracy of already existing codes (Bajwa et al. 2021, Hanna et al. 2025). With an accuracy level of eighty-five percent in unlabeled tests from laboratories and 96 percent in randomly selected already labeled laboratory tests, their tree-based AI algorithm effectively identified the LOINC code. Manual assessment demonstrated that the machine learning method was accurate eighty-three percent of the time when it diverged from the presently assigned LOINC code, yet the previously assigned label's rate of precision was 72 percent (Rabbani et al. 2022). Islam et al. created a deep learning machine learning model that may suggest which laboratory tests a provider should request in the same field of clinical decision support and laboratory utilization. Instead of forecasting particular test outcomes, their algorithm forecasted the initial order of exams. With an AUROCmacro of 0.76 and an AUROCmicro of 0.87, their neural network model demonstrated a moderate level of performance when using characteristics such as clinical diagnoses, medications, previous laboratory tests, and demographic data. The algorithm was trained using previous testing patterns, but no expert assessment was conducted to determine whether these prior ordering habits were optimal in the first place (Bonini et al. 2002). This is a significant weakness of the study. In contrast to a ground truth of fractionated LDL-C measurement, Lee et al. from South Korea presented a neural network deep learning model to predict low-density lipoprotein cholesterol (LDL-C) from high-density lipoprotein cholesterol (HDL-C), total cholesterol, and triglycerides. They demonstrated that, with a root mean squared error of 8.1 mg/dL as opposed to 10.8 mg/dL and 8.3 mg/dL, respectively, their model outperformed the old Friedewald equation and Martin's "novel method." Lastly, Dunn et al. completed a research study that used data from wearable devices like accelerometers and electrodermal probe sensors to predict typical laboratory tests using a machine learning regression model. Regretfully, meaningful performance was not achieved by this futuristic approach to laboratory test prediction. For instance, the laboratory test for which the model scored best, the fluctuation in hematocrit level, could only be explained by 21% of the wearable data included in their model (Yu et al. 2018b).

8. AI in Biochemical Analysis

By enabling early disease detection, individual risk assessment, and treatment result forecasting, AI has revolutionized disease diagnosis and management. Early detection and improved outcomes for patients are made possible by ML algorithms that can evaluate clinical and laboratory data to find biomarkers linked to particular diseases. In addition, medical professionals may forecast disease progression, improve methods of treatment, and reduce adverse consequences with the help of AI-driven predictive modeling. In the advanced stages, it will lead to improved patient treatment and decrease the cost burden. In a previous study, authors assessed the association of Vitamin D response and cognition markers by artificial neural network

(ANN) methods, which help in accurate prediction of responsiveness to vitamin D treatment that ultimately helps patients in their treatment (Morrow et al. 2007). A study in the USA on a seven-metabolite panel showed promise as a trustworthy kidney cancer screening tool by identifying RCC status with good accuracy, sensitivity, and specificity. By combining mass spectroscopy and computer learning, scientists in Brazil developed an efficient platform for COVID-19 recognition in plasma specimens. Strong diagnostic potential was demonstrated by the model, and biological marker comparison enhanced its robustness and appropriateness for actual screening. The capacity of AI to integrate multiple omics layers—genomics, transcriptomics, proteomics, and metabolomics—into one structure for physiological prediction can be seen in the right panel. The following illustration shows how AI improves the structure of enzymatic variants and uses integrative analysis for evaluating intricate biological processes (Ashraf et al. 2025b). AI in omics and enzyme engineering: Since it makes predictive modeling, integrative analysis, and high-throughput interpretation of data possible, AI is quickly revolutionizing omics research and enzyme engineering. AI models like decision forests and deep neural networks are used in enzyme engineering to forecast the functional implications of amino acid substitutions, maximize catalytic effectiveness, and generate enhanced variations with unique activity. The World Health Organization (WHO) released its governance and ethics principles to tackle the increasing dangers of AI misuse in biochemical research. These suggestions support the ethical creation of AI through putting a strong emphasis on transparency, security, and worldwide equity in application and access (You et al. 2025).

9. AI in Biochemical Diagnosis

Convolutional neural networks (CNN) and transformers are instances of algorithms for DL that have outperformed conventional diagnostic approaches in accuracy and efficiency, particularly for neuroimaging for diseases like AD. AI-powered strategies additionally allow personalized healthcare through wearable technology integration, omics-based system biology, and real-time health monitoring (Lee and Durant 2022). In order to get around these challenges, this work highlights solutions like global cooperation, free software platforms, and transparent AI. The transdisciplinary scope and future potential of AI in biochemistry are further demonstrated by novel uses such as AI-enhanced microplastic toxicity, sleep biochemistry, herbal part modeling, and microbiota and host interaction mapping.

10. Discussion

10.1 Obstacles on the Field

Scientifically speaking, the accuracy of data is one of the primary hurdles that artificial intelligence algorithms in medical applications must overcome (Panch et al. 2018). The highest possible performance any algorithm may achieve has been limited by the challenge of missing or incorrectly categorized data in laboratory information systems. The cost of building a computer system and recruiting employees who have the necessary computational expertise to develop, deploy, maintain, and upgrade machine learning techniques and the software tools required for running them is another technical as well as financial obstacle. Clinically speaking, artificial intelligence in the laboratory sciences has to be standardized and strictly controlled considering that it is still an area that is relatively fresh. At this moment, there are no standards for the most effective approaches to validate the use of machine learning systems in medical applications. There are no recognized standards for laboratory-developed programs or for the validation of vendor-developed software, even in pathology disciplines like digital pathology, where clinical machine learning methodologies are changing quickly (Darrow et al. 2021, Kocak 2024). Indeed, a group was subsequently formed by the College of American Pathologists (CAP) with the goal of filling this gap, which involves establishing standards for laboratories for artificial intelligence (AI) applications. In the same way, oversight organizations like the Food and Drug Administration have not yet entirely settled whether or not they intend to regulate AI applications created in laboratories. The Food and Drug Administration (FDA) released an action plan in 2021 to update its proposed regulatory structure for software based on learning algorithms and artificial intelligence as a medical device. This is an essential phase towards controlling the healthcare software business. It remains unknown, nonetheless, how state and federal authorities will react regarding machine learning strategies created in laboratories (Bajwa et al. 2021, Hanna et al. 2025). Subsequently, for medical laboratory experts to thoroughly adopt this innovation, additional studies analyzing the practical deployment of these applications in laboratory settings and demonstrating their reliability and security will be required.

11. Thoughts and Prospects for the Future

A system based on rules or human experts might not be capable of deducing complicated linkages and correlations from enormous quantities of data; however, machine learning can. Likewise, machine learning can uncover novel trends and applications and continually use new information to enhance its effectiveness, whereas stationary rule-based computer programs rely on previously established knowledge. Consequently, artificial intelligence's effectiveness for automation has been considered one of the most intriguing characteristics in laboratory medicine. The examined work demonstrates significant advantages over the current rule-based methods of selecting specimens for manual validation or verification by using predictive ML algorithms. These approaches are currently demonstrating outstanding efficiency, and we believe they will soon be standard in contemporary labs. The ability of ML to use enormous amounts of historical health information to generate more tailored results from test interpretation is another fascinating use scenario. Some researchers suggest an incredibly straightforward technique for leveraging diagnosis codes to create slightly more customized benchmark ranges. In order to provide individualized laboratory test reference ranges and promote precision diagnostics, we expect additional studies into deeper algorithms that take into consideration the patient's whole clinical situation. Lastly, even though the subject has been addressed

CHAPTER 15

Automated Patient Monitoring and Decision Support in Intensive Care Units (ICUs)

Adebayo, Paul Olujide,[1] *Adesina, Ayodele Joshua,*[2] *Usman-Hamza Fatimah Enehezei,*[1] *Abikoye, Oluwakemi Christiana,*[1] *Oladipo, Idowu Dauda,*[1] *Awotunde, Joseph Bamidele*[1,*] and *Abdulraheem, Muyideen*[1]

1. Introduction

Intensive Care Units (ICUs) are specialized units of hospitals that are created to offer holistic and extensive care to patients with severe or life-threatening conditions and injuries. The units have advanced medical technologies and multidisciplinary teams of medical specialists such as intensivists, nurses, and respiratory therapists, among other specialists. The main objective of ICUs is to stabilize patients, monitor vital signs, and provide life-sustaining care. Since the conditions in ICUs that are addressed are complex and urgent (organ failure, severe infections, and post-operative complications), survival and recovery require accurate and timely monitoring. ICUs are also areas of innovation in clinical care (Smith et al. 2020), as new technologies and protocols are usually enhanced and tested on a smaller scale before they become widely applicable. It is one of the most challenging and yet important spheres of modern medicine, as it is fast-paced and requires constant attention.

[1] Department of Computer Science, University of Ilorin, Ilorin, Nigeria.
Emails: adebayo.po@unilorin.edu.ng, https://orcid.org/0000-0002-4088-0634; usman-hamza.fe@unilorin.edu.ng, https://orcid.org/0009-0004-1115-4555; abikoye.o@unilorin.edu.ng, https://orcid.org/0000-0002-8912-6333; odidowu@unilorin.edu.ng, https://orcid.org/0000-0002-1472-2237; muyideen@unilorin.edu.ng, https://orcid.org/0000-0003-4568-7566
[2] Department of Computer Engineering, University of Ilorin, Ilorin, Nigeria.
Email: adesina.ja@unilorin.edu.ng, https://orcid.org/0009-0003-0102-4426
* Corresponding author: awotunde.jb@unilorin.edu.ng, https://orcid.org/0000-0002-1020-4432

DOI: 10.1201/9781003783855-19

The traditional systems are usually quite limiting, even though monitoring is extremely vital in the ICUs. Traditional monitoring is based on the use of standalone sensors that monitor the specific physiological parameters of the individual, including heart rate, blood pressure, oxygen saturation, and respiratory rate. They tend to be utilized as disjointed enclosures and not interconnected or interoperable, which might be counterproductive to evaluate the state of a patient comprehensively. Moreover, oncologists have a huge volume of generated data in ICUs, enough to saturate their minds and trigger alarm fatigue, i.e., a state when the constant use of false alarms in ICUs becomes desensitizing and slows responses to actual crises. The impact of such challenges does not just ruin patient safety, but also results in burnout and low efficiency among clinicians in the provision of care. It makes the integration of monitoring systems even more complex because there are no standard forms of data or specifications of communication, and it is also difficult to present a single picture of patient health. To eliminate these issues, healthcare systems are quickly turning to automation and decision support technologies. Biosensors, the Internet of Things (IoT), and real-time data help automate patient-monitoring systems as well as analytics (Brown 2019) to continuously read and handle patient data. The systems reduce the amount of data entry and errors made by hand and permit preventative actions.

Decision Support Systems (DSS), particularly Clinical Decision Support Systems (CDSS), aid clinical judgment and can provide evidence-based suggestions, predictive warnings, and diagnostic assistance (Lee and Kumar 2021). These tools help achieve test accuracy and therapeutic outcomes as they can identify minor trends in patient data that the human eye cannot detect using Artificial Intelligence (AI) and machine learning (ML). This is a paradigm shift in ICU care—the shift toward anticipatory medicine.

They also facilitate better resource allocation and effective workflow that supports clinicians to undertake emergency tasks.

This chapter aims to determine the change effect of automated monitoring and decision support systems in the ICU setting. It begins with an examination of the principle of patient monitoring, the major physiological parameters, as well as the disadvantages of traditional equipment. The future direction lies in the seamless integration of automation technologies, with a strong emphasis on incorporating advanced biosensor systems, IoT systems, and electronic health records (EHRs). The role of decision support systems is analyzed accordingly, with the rule-based model and AI-driven model taken into account, and how they are applied in circumstances of specific significance, such as the recognition of sepsis and the control of ventilation. Other technological enablers include big data analytics, cloud computing, and robotics. It aims to provide in-depth insight into how automation and decision support can be used in ICUs to improve patient outcomes as well as operational effectiveness. Information about the benefits, opportunities, and future of these innovations is discussed in this chapter, with respect to how smart systems are reshaping the discipline of critical care in ICUs.

2. Foundations of Patient Monitoring in ICUs

Patient monitoring in ICUs is crucial for critical care, enabling timely detection of physiological changes and supporting life-saving interventions (Khan 2025). Continuous monitoring provides real-time data on vital parameters like heart rate, blood pressure, oxygen saturation, respiratory rate, and temperature, acting as an early warning system to intervene before complications escalate (Selvaraju et al. 2022). Non-invasive and invasive are the two main techniques used in ICU monitoring. In non-invasive methods, without penetrating the body, vital signs are assessed, providing real-time clinical data. Common non-invasive methods are surface thermometry, pulse oximetry, blood pressure cuffs, capnography, and ECG. Invasive monitoring, on the other hand, inserts devices into the body to measure critical functions (Jung et al. 2020), which include blood pressure, fluid status, cardiac output, and intracranial pressure. These invasive techniques carry infection and complication risks, but are highly accurate.

Modern ICU integration uses advanced monitoring systems for data consolidation from multiple devices into a central platform (Awotunde et al. 2022). This helps clinicians identify trends, improve decision-making, and reduce error risk. These systems have in-built alarms and alerts to guide staff in prioritizing urgent changes. Data interpretation is another crucial foundation in ICUs. In each patient's condition, accurate measurement, technological integration, and skilled interpretation are needed to understand the significance of data. Thus, critically ill patients in high-risk conditions are ultimately provided with personalized care, improved survival rates, and precise and rapid outcomes.

2.1 Key Physiological Parameters Monitored in ICU

In ICUs, monitoring key physiological parameters is crucial in identifying early signs of deterioration and ensuring quick interventions (Stellpflug et al. 2021). Respiratory rate, blood pressure, oxygen saturation, heart rate, and temperature are some of the parameters that provide vital organ function snapshots. Advanced monitoring tracks conditions such as intracranial pressure, arterial blood gases, or central venous pressure when necessary. These provide medical practitioners with cardiovascular, respiratory, and metabolic stability measurements to ensure appropriate treatment decisions for patient recovery and survival. Some major physiological parameters monitored are shown in Table 1.

2.2 Conventional Monitoring Tools/Devices

The ICU traditional monitoring equipment is used during the routine monitoring of patients. These tools, including respiratory monitors, ECG machines, pulse oximeters, thermometers, and blood pressure cuffs, are commonly used to monitor vital parameters such as heart rate, blood pressure, oxygen saturation, temperature, and respiratory rate. Limitations notwithstanding, the devices are relied upon to provide the required information and support clinical decision-making during critical care and in general healthcare facilities.

Table 1: Physiological parameters monitored in patients.

Physiological Parameter	**Description**	**Monitor**
Heart Rate (HR)	Heart rate (HR) is the number of times the heart beats per minute. In ICUs, it reflects cardiac function and can indicate stress, arrhythmias, or hemodynamic instability.	ECG or bedside monitors
Blood Pressure (BP)	Blood pressure (BP) is the force exerted by circulating blood on artery walls. In ICUs, it indicates cardiovascular stability and guides fluid management, vasopressor use, and shock detection.	Non-invasive (cuff) or invasive (arterial line)
Respiratory Rate (RR)	Respiratory rate (RR) is the number of breaths taken per minute. In ICUs, it helps assess ventilation, detect respiratory distress, monitor treatment response, and identify early signs of deterioration.	Spontaneous or ventilated breathing
Oxygen Saturation (SpO_2)	Oxygen saturation (SpO_2) measures the percentage of hemoglobin bound with oxygen in the blood. In ICUs, it indicates oxygenation status, guiding respiratory support and detecting hypoxemia or respiratory failure.	Using pulse oximetry
Body Temperature	Body temperature reflects the balance between heat production and loss in the body. In ICUs, it helps detect infection, inflammation, hypothermia, or hyperthermia, guiding treatment and monitoring metabolic stability.	Core or surface measurement
Electrocardiogram (ECG)	An electrocardiogram (ECG) records the heart's electrical activity. In ICUs, it helps monitor heart rhythm, detect arrhythmias, ischemia, or conduction abnormalities, and guide timely interventions for critically ill patients.	Cardiac rhythm and electrical activity
Central Venous Pressure (CVP)	Central Venous Pressure (CVP) measures the pressure in the thoracic vena cava near the right atrium. In ICUs, it guides fluid management, assesses venous return, and evaluates cardiac function.	Fluid status and venous return
Arterial Blood Gases (ABGs)	Arterial Blood Gases (ABGs) analyze oxygen, carbon dioxide, and blood pH levels. In ICUs, they assess respiratory function, acid-base balance, and guide ventilation, oxygen therapy, and metabolic management.	Oxygenation, ventilation, acid-base balance
Intracranial Pressure (ICP)	Intracranial Pressure (ICP) measures the pressure within the skull from brain tissue, blood, or cerebrospinal fluid. It is used in ICUs, in the management of head injuries, strokes, and prevention of secondary brain damage.	In patients with brain injury
Urine Output	Kidney and general fluid balance is reflected by urine output. It can be used as one of the indicators of renal perfusion, shock, and response to fluid therapy or ICUs for critical illness management.	An indirect marker of kidney function and perfusion

3. Automation of ICU Monitoring

The evolution of the automated system of patient monitoring transformed the manner of providing healthcare services and discarded the barbaric machines in favor of experimental machines that have AI enhancements, which enable remote care services in real time. This process started in the early 20th century when simple telecommunication technologies such as the telephone and radio were invented, which formed the foundation of the transmission of medical data over a distance. By the middle of the twentieth century, there was the appearance of telehealth that enabled doctors to consult remotely with the help of radiology pictures and neurological exams provided remotely. The 1960s became a turning point as space exploration was an incentive to advance innovations in biomedical telemetry. The necessity of NASA to remotely monitor the vital signs of astronauts resulted in the creation of remote wear sensors and telemetry systems—technologies which would be reconfigured to be used in civilian medicine. Hospitals started to install heart rate, blood pressure, and oxygen saturation bedside monitors in the decades that followed, which allowed critical patients to be constantly monitored.

Remote Patient Monitoring (RPM) was a fashion of the 2000s and brought care out of the hospital. RPM systems were wearable and home-made sensors that input health data, and it was sent to the provider to analyze. It was particularly beneficial in managing chronic disease treatment like diabetes and hypertension, hospital readmissions, and patient outcomes. The past few years have seen Artificial Intelligence (AI) transform RPM. Smart AI systems have the capacity to handle unlimited volumes of patient data through real-time delivery of predictive and tailored care plans. The latter are an amalgamation of machine learning algorithms, relevant wearable devices, mobile applications, and even smart home devices. The intention behind their introduction is to monitor vital signs, medication adherence, and even any indications of mental illness. To provide an example, AI-based RPM can indicate the signs of heart pain or sugar level changes at the right time to implement measures so that the situation can be avoided, a contactless solution to problems. In response to the need to find no-contact solutions and in the wake of the COVID-19 pandemic, there was an increased chance of employing automated monitoring. RPM is nowadays not a luxury but modern treatment. Regulatory authorities have also followed suit, and new frameworks have been updated to facilitate the implementation of remote monitoring services more broadly, as well as reimbursement thereof (Shaik et al. 2023). In the future, AI, Internet of Things (IoT), and 5G connectivity are going to be merged, which will enable even more responsive and scalable monitoring systems. These technologies will also allow the free flow of data, diagnostics will be faster, and equitable access to care will be provided particularly in underserved areas. Basically, passive monitoring has been replaced by active and smart care (mechanized patient monitoring). It is no longer a question of watching it but rather customizing healthcare interventions in ways that were never imagined. Combined with biosensors, the Internet of Things (IoT), and the real-time compilation of information are redefining the Intensive Care Unit (ICU) monitoring. This has resulted in a greater volume of accuracy, timeliness, and predictive data in an Intensive Care Unit than ever before.

This is a change that occurs in the biosensors. These analysis tools detect physiological changes by converting biological interactions into electrical impulses. The modern biosensors can continuously monitor vital signs such as heart rate, sugar level in blood, oxygen concentration, and even an indicator of an organ or biochemical infection (Hosain et al. 2024). They are perfect in ICU settings for enabling 24-hour focused attention where they are required for them to be non-invasive and miniaturized.

The IoT appliances are the connective tissue among the biosensors and the clinical decision-making systems. For the seamless transmission of data and remote monitoring, considering minimal latency, IoT connects sensors to cloud-based applications. The connectivity can enable clinicians to monitor patient status in real time and get alerts when readings are abnormal, and historical data can be accessed to analyse trends. Smart biosensors combined with the IoT have made remote ICU monitoring a possibility and helped ease the workload on personnel and increased patient safety.

The last element in the puzzle is real-time data collection and Artificial Intelligence (AI)-powered dashboards. Aakala (2025) showed that AI dashboards in ICUs can combine data of multiple sensors to predict clinical deterioration prior to the process. Such systems employ machine learning models including convolutional neural networks (CNNs) and long short-term memory (LSTM) to identify slight changes in patient vital regions and provide proactive warnings. It was stated that the research has reduced the number of critical incidents by 25% and shortened the response time by 30%, which is a demonstration of the life-saving power of this technology.

The closed-loop system of constant monitoring, predictive analytics, and timely interventions is achieved by leveraging all these technologies. Unlike the usual telephony systems, which react to occurrences in the future, the combined approach has the capability to anticipate complications and allows clinicians to be proactive. It also creates personalized medicine whereby an intervention can be formulated based on patient peculiarities. This combination in a closed-loop system of biosensors, IoT, and real-time analytics is changing the nature of care in an ICU. It enhances situational awareness, risk events, and patient outcomes. These technologies are also becoming more ubiquitous and developing more and more, and with increasing availability they will become more mainstreamed in the ways they will be utilized to bring a new wave of smart critical care.

The interoperability principle is the fluency and comprehension of health data that has been captured among systems, devices, and platforms. Interoperability implies the automatic capture of data in the ICU and its transfer to the EHR that are performed by the use of the instruments that monitor the patient like ventilators, infusion pumps, and heart monitors. This removes manual charting, will lower the errors in transcription, and will allow clinicians to get a real-time status of the patient. Although there are widespread digitizations, the world is still yet to achieve real interoperability.

Many ICU systems still rely on fragmented data silos, where devices and EHRs operate independently. This disconnect can delay interventions and compromise care quality. A major hurdle is the inconsistent implementation of standards like

FHIR which, while powerful, vendors often adopt partially or differently, leading to semantic mismatches. Moreover, "information blocking" by vendors through opaque APIs or licensing fees can restrict data flow, undermining the promise of connected care. Regulatory efforts like the 21st Century Cures Act aim to curb this, but business incentives often favor data hoarding over collaboration (Preston 2022).

In 2025, several innovations are bridging these gaps. AI-driven data normalization tools are helping harmonize disparate data formats, making device outputs usable across EHR platforms (Jimenez-Pastor et al. 2025, Goyal and Fiorini 2025). The Trusted Exchange Framework and Common Agreement (TEFCA) initiative is also gaining traction, offering a standardized framework for nationwide health information exchange (Adler-Milstein et al. 2023). Medical device integration platforms now support bi-directional, real-time data exchange, allowing ICU monitors to not only send data to EHRs but also receive alerts and treatment protocols. This creates a feedback loop that enhances clinical responsiveness and supports personalized care plans. Interoperability transforms ICU monitoring from a reactive to a proactive model. As healthcare systems continue to evolve, the ability to connect devices and EHRs meaningfully will define the next frontier of critical care.

Automated monitoring systems in ICUs have revolutionized critical care by enhancing real-time surveillance, reducing human error, and improving patient outcomes. These systems integrate technologies such as AI, IoT, and machine learning to continuously track vital signs and alert clinicians to potential emergencies. The following are some of the recent case scenarios that demonstrate how they are applied in practice. Aakala (2025) presented the use of AI-based dashboards in ICU facilities. They are dashboards that consolidate data of various sensors tracking heart rate, blood pressure, oxygen saturation, among others. The system anticipates adverse events and predicts them using CNNs and LSTM models. This system helped to decrease critical incidents by 25% and clinician response time by 30 in pilot testing. The predictive alerts enabled the medical staff to act earlier, which greatly enhanced patient outcomes. The cases bring out the importance of AI in converting raw data into a digital dashboard for actionable insights, through which proactive care becomes possible.

Raju et al. (2023) outlined an Indian-developed ICU monitoring system based on IoT. This system uses a system of smart sensors that continually monitor vital signs that comprise heart rate, respiratory rate, and oxygen levels. The information will be transmitted wirelessly to an analysis centre where it will be analyzed and any anomalies reported. The system also includes a user-friendly dashboard and multiple tiers of security as the means of ensuring integrity and confidentiality of the data. According to these authors, this system reduced the inaccuracy of information input through manual processes, and it became quicker in responding to life-endangering events. It will be involved in a pilot implementation which will evaluate its efficiency and patient ICU care efficacy (Vellela et al. 2024).

In Ecuador, researchers at Guayaquil Hospital designed a mobile-built intensive care monitoring system that includes a mobile application, a database server, and a transactional server (Collaguazo et al. 2022). The application displays the real-time patient data and offers emergency alarms within 3 seconds, in case an emergency occurs. With this system, physicians are able to provide follow-up on patients even when they do not sit in the ICU and to exercise appropriate actions even when not in the ICU.

The researchers concluded that timely and accurate alerts dramatically lowered response phase in emergency cases and also enhanced the recovery rate. The case illustrates the importance of mobile integration as a solution in ensuring that the ICU monitoring services are stretched beyond the physical location of the unit. These case scenarios outline possibilities that automated monitoring systems may have in ICUs. These technologies may be applied to offer round-the-clock, proactive surveillance, independent of whether it is the AI dashboards, sensor networks, or just mobiles and cells, and enhance clinical decision-making and patient security. The medical care environment is gradually embracing digital innovation, and as such, these systems will become standard in critical care environments soon.

4. Decision Support Systems (DSS) in Intensive Care Units

ICUs are special wards in hospitals which are dedicated towards offering round-the-clock services to patients who are critically ill or have life-threatening conditions. These hospitals have electronic health record surveillance devices in these wards, and these devices also monitor the vitals of the patients. In an effort to manage the high volumes of health data which these devices operate and enable timely healthcare intervention, decision support systems are used to aid clinical practitioners by using rule-based, AI-driven, and predictive analysis systems to process data and trends to enhance informed clinical decision-making.

A Clinical Decision Support System (CDSS) is a computer application that supplements medical judgments by acting on medical information, algorithms, and patient records to offer actionable information to enhance clinical care delivery (Elhaddad and Hamam 2024). They can be combined with electronic medical records to reveal information when analysing large and complex health data that may otherwise be overlooked when interpreted by humans, hence complementing clinical decision-making (Sutton et al. 2020). These systems are designed to aid medical health practitioners in making critical decisions as opposed to substituting them, in contrast to the opposite opinions. Time is paramount in such situations and accelerated decision-making is informed with accelerated data processing such CDSS offer.

The analysis of the clinical data is performed with the help of CDSSs, which generate automated warning signs in case of a violation of vital signals through different predictive algorithms to identify signs of warning (Grechuta et al. 2024). They have also been very effective in providing real-time surveillance and alerts.

As highlighted by Olanrewaju et al. (2024) during the COVID-19 pandemic, support systems reshaped the way healthcare professionals examined critically ill patients in the ICUs. Due to the severe impact of the pandemic, it was crucial to identify conditions that could quickly worsen, and the quick intervention provided by CDSS led to improved patient outcomes. The incorporation of CDSS with Computerized Provider Order Entry (CPOE) systems has helped to significantly improve the performance of health practitioners, reduce medication errors, and ultimately improve patient outcomes (Shahmoradi et al. 2021). A study that involved definitions such as postoperative hemorrhage and anaphylactic shock. On the same note, a light gradient boosting model was used to create an AI-CDSS based on blood count with differential data. This system facilitated clean predictions and established that these systems have the capacity to detect sepsis correctly without massive clinical parameters, hence enhancing patient care (Lin et al. 2025).

Early withdrawal of mechanical ventilation in ventilator management or increased duration of mechanical ventilation is associated with increased mortality (Lin et al. 2025). To determine the most appropriate ventilation timing, Prasad et al. (2017) created a decision support system based on reinforcement learning through Q-iteration to learn the weaning policy with the help of historical data of the ICU. The algorithm was promising because it could ascertain whether a patient was to be extubated or not depending on his or her vitals, which indicated it was fit to make clinical decisions.

Clinicians have employed knowledge-based CDSS and AI-CDSS to provide their patients with recommendations on the basis of clinical guidelines, which minimizes the extreme outcomes that can arise due to the bad choice of treatment and prescription (Armando et al. 2023). These systems have contributed to better personalized healthcare through the provision of quick and precise medical condition diagnosis. Segal et al. (2019) integrated a system into the electronic medical record using outlier detection algorithms to monitor the effectiveness and validity of administered prescriptions. Results of the alert system demonstrated that the majority of the prescriptions were clinically valid, with physicians responding to generated alerts in real time.

In situations where untimely detection of a health condition could significantly impact a patient's well-being, it is imperative to develop warning systems that flag risk factors as timely as possible. Machine learning algorithms have been integrated to build CDSSs that detect the early occurrence of chronic diseases such as cancer, diabetes, respiratory diseases, and cardiovascular diseases, which remain the leading causes of mortality. These interventions constitute early warning signals that improve patient outcomes. In intensive care, medical decisions are usually high stakes as any oversight from clinicians could have severe consequences. This is the major reason doctors approach AI support systems with caution despite the advancement of algorithms over the years. The black box nature of these algorithms establishes a level of hesitation to fully embrace AI-based decisions (Zhang et al. 2024).

While understanding that CDSS holds a lot of promise for improved healthcare systems, these systems are developed to support critical decision-making from experts rather than replace these experts. Therefore, the development of these systems should integrate expert knowledge. Cuvelier et al. (2021) examined the role clinical pharmacists play in the development of a decision support rule for generating pharmaceutical intervention alerts. By establishing a three-step evaluation, it was observed that the involvement of pharmacists deemed only 24.6% of 1,430 generated alerts relevant. It is important for a level of cross-functional collaboration to exist between experts and clinical decision support systems to enable better trust in their usage to improve outcomes.

5. Technological Enablers

Transforming ICUs into innovative, digital spaces relies on a range of technologies; these technologies go beyond traditional monitoring devices. As patient care becomes increasingly focused on data, new technologies such as artificial intelligence, big data, advanced computing systems, and robotics aid in monitoring and informed decision-making. These technologies work together to improve healthcare. They not only increase the capacity of medical workers and contribute to the rapid decision-making process but eventually achieve better patient safety and health. Largely, this can be explained by the fact that numerous students face challenges related to their family background (Bignami et al. 2023, Ongenae et al. 2015, Vogt et al. 2019).

5.1 Artificial Intelligence (AI) and Machine Learning (ML) in the Patient Monitoring

Wearable devices contain AI and ML that monitor the health conditions of patients, particularly in ICUs, which is an important part of the healthcare system (Coffman 2018, Nahar et al. 2020, Olujide et al. 2025, Srinivasan et al. 2023). The combination of AI and ML has improved healthcare services by offering immediate data processing and recommendations on the health conditions of a patient. This innovation does not only assist healthcare professionals but also leads to fast decision-making. Prior to the development of AI and ML, the old systems were enhanced to monitor the vital signs in patients, including blood pressure, oxygen levels, and heart rate. These provide the warning signs that show that a patient is deteriorating. AI algorithms are created to examine long-term health data, identify the trends in such data, and make certain predictions (Ramesh et al. 2023). For example, AI systems may notify healthcare providers about possible problems in a patient based on the monitoring feedback, which can be promptly addressed to make a major improvement in patient outcomes. Using improved machine learning tools, medical professionals are able to sharpen their skills in identifying abnormalities and anticipating a health risk, which eventually changes their work with patients and their management (Gupta 2024).

The machine learning solution thereby enables the clinical staff to intervene in the health condition earlier, before it escalates. These proactive steps demonstrate that AI is not just a tool; it actively helps improve care in the ICU, leading to better patient outcomes.

Moreover, Table 3 highlights the practical, clinical, and technological relevance of AI and ML in patient monitoring. An example is offered across the ICU, wearable devices, and chronic disease management.

5.2 Big Data Analytics and Real-Time Stream Processing

Big Data Analytics involves examining large, complex datasets to identify patterns and insights that enable informed decision-making. Real-time stream processing continuously analyzes and processes data as it is generated, allowing immediate insights and timely responses to events (Sadiku et al. 2020). The ICU is an

Table 3: ICU, Wearable devices, and Chronic Disease Management Patient Monitoring with AI and ML.

Domain	**Role of AI**	**Role of ML**	**Examples/Applications**
ICU Monitoring	Automates continuous analysis of vital signs and imaging to detect critical changes.	Learns from historical ICU patient data to predict sepsis, multi-organ failure, or cardiac arrest.	AI-driven sepsis prediction systems; ML-based models predicting mortality risk in critical care.
Wearable Devices	Processes large streams of biosensor data (ECG, SpO_2, glucose monitors) in real time.	Learns individual patient baselines to distinguish between normal fluctuations and early signs of deterioration.	Smartwatches detecting arrhythmias; ML-powered glucose monitoring predicting hypoglycemia
Chronic Disease Management	Provides personalized health recommendations and automated alerts for long-term monitoring and care.	Trains predictive models to assess risk progression in chronic conditions (e.g., diabetes, hypertension)	AI-enabled apps for hypertension monitoring; ML models predicting diabetic complications.
Early Diagnosis & Risk Stratification	Utilizes intelligent algorithms to identify subtle patterns that humans often overlook.	Learns risk profiles and stratifies patients based on the probability of disease progression.	AI-assisted ECG interpretation; ML-based cancer recurrence prediction.
Clinical Decision Support	Generates actionable alerts, drug dosage recommendations, and treatment pathways.	Reduces false alarms by continuously refining predictions from new patient data.	AI-based ICU decision-support dashboards; ML models optimizing ventilator settings.
Remote/ Telehealth Monitoring	Analyzes data streams from home-based devices to support remote consultations and patient care.	Builds adaptive models that improve over time with continuous patient feedback.	AI-enabled remote COVID-19 monitoring; ML algorithms tracking rehabilitation progress.
Data Integration & Analytics	Fuses multimodal data (lab results, imaging, vitals, genomics) for holistic monitoring.	Finds predictive associations between diverse datasets to enhance prognostic accuracy.	AI-driven hospital command centers; ML-based predictive analytics in electronic health records (EHRs).

environment where every patient generates massive volumes of heterogeneous data from monitors, ventilators, infusion pumps, imaging systems, and EHRs (Defense et al. 2022, Luo et al. 2022). Big data analytics transforms this overwhelming data deluge into actionable intelligence. By aggregating structured and unstructured datasets across temporal and spatial scales, analytics platforms enable holistic assessments of patient status.

A practical example is the **MIMIC-IV database** (Medical Information Mart for Intensive Care), one of the most enormous open-access datasets for critical care (Khope and Elias 2022). Researchers worldwide use MIMIC to develop and validate predictive models, including algorithms for mortality risk and ventilator management. Beyond research, real-time stream processing platforms such as **Cerner's St. John Sepsis Agent** monitor continuous data feeds to detect sepsis onset, generating alerts that have been shown to reduce mortality rates. Such examples demonstrate how analytics and streaming systems not only organize data but also deliver life-saving insights within clinically meaningful timeframes (Callcut et al. 2021, Liu et al. 2022).

5.3 Cloud Computing, Edge Computing, and IoT Frameworks in ICUs

Cloud computing provides on-demand access to computing resources over the internet. The on-demand access to computing resources (servers, storage, databases, software) is available over the internet on demand, offering scalability and flexibility. Likewise, Edge Computing is a distributed computing model that processes data closer to its source (e.g., sensors, devices) to reduce latency and bandwidth use. The Internet of Things (IoT) is a network of interconnected devices embedded with sensors and software that collect, share, and act on data over the internet (Kaur 2020, Rani and Sabri 2020).

Modern frameworks in ICUs could be the integration of cloud computing, edge computing, and IoT devices. The combination enables the creation of a digital system to collect, process, and store ICU environment data. For instance, cloud services such as Microsoft Azure for Healthcare (Lu et al. 2013) and Amazon Web Services (AWS) HealthLake (Sousa et al. 2022) enable healthcare organizations to store and process large amounts of data efficiently. These sample platforms are designed to support learning across different platforms and help develop a general AI model. It also enhances the reliability of algorithms and mitigates biases inherent in data from a single ICU.

In the event of a framework failure, edge computing brings computation closer to the patient's bedside, thereby reducing latency. This design ensures uninterrupted monitoring of events for prompt action. For example, **GE Healthcare's Edison platform** integrates edge and cloud computing to deliver AI-powered insights from healthcare data stored (Idelchik and Kogan 2012). This solution enables ICU teams to monitor patients in real time. Similarly, an IoT framework integrates various monitoring devices. This design enhances communication and facilitates the coordination of patient care. This system is scalable and resilient, which are essential qualities in critical healthcare settings.

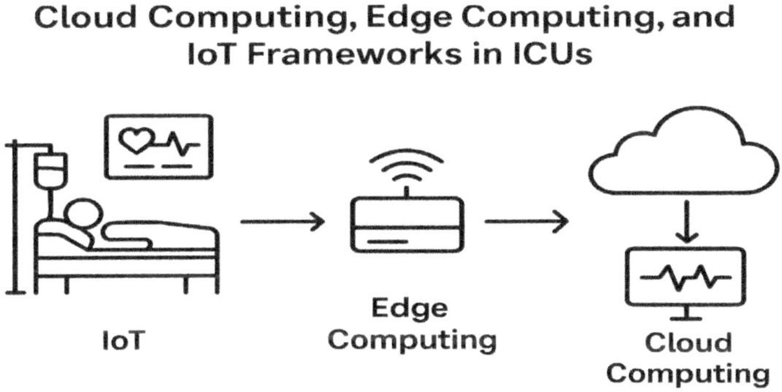

Figure 2: Cloud computing, Edge computing, and IoT Frameworks in ICUs.

Table 5: Description and application of various aspects of patient safety in ICU.

Aspect	Description	Example in ICU
Patient Safety	Continuous monitoring reduces the likelihood of missed critical events, lowers the risk of adverse incidents, and ensures faster response times.	Automated alert systems detecting sudden oxygen desaturation.
Clinical Outcomes	Supports timely, evidence-based interventions, leading to reduced mortality, fewer complications, and shorter ICU stays.	AI-assisted sepsis detection enables earlier treatment and improved survival rates.
Length of Stay	Optimized treatment pathways shorten ICU admission times.	Automated decision support guiding ventilator weaning.

the quality of patient care in the ICU. Using a decision support system on a large scale can help reduce deaths and illnesses. It can also lead to quicker treatments. As an example, the use of automated protocols reduced complications and shortened the length of stay on the ventilator. These assist patients in being weaned off the ventilators. Notably, automation collaborates with clinical skills. It improves the decision-making by the doctors, as well as results in safer and more effective care.

6.2 Reduction in Human Errors and Workload for Healthcare Professionals

The experience of working in an ICU is challenging for healthcare workers. They must be smart, handle a lot at the same time, and work under pressure. The task may cause those staff on duty to make errors. Nevertheless, automated and decision support systems may be implemented into ICU workflows to address these issues and achieve better patient outcomes. Computerized reminders can be used to detect potential drug interactions, medication errors, or oversights. These notifications enhance security. The systems enable clinicians to concentrate more on making complex decisions and direct patient care by eliminating repetitive and error-prone tasks. Such workload adjustments assist in minimizing errors and foster personnel welfare and retention by reducing mechanical and non–value-added work.

6.3 Early Prevention of Worsening and Anticipatory Intervention

Anticipatory care is possible through automated monitoring and decision support systems, which can detect minor physiological changes before they escalate into critical events, as compared to traditional methods, which detect deterioration once symptoms manifest. ML early warning systems can predict (several hours before) the occurrence of septic shock and acute respiratory distress. These messages give clinicians the urgent opportunity to act to alter the ICU practice to a more proactive model promptly to prevent more injuries and deaths.

6.4 Optimized Resource Allocation in ICUs

In addition to the improvement of patient benefits, automated monitoring and DSS enhance the clinical management of resources at the system level. The researchers can distribute limited resources more effectively, such as ventilators and specialist staff, the most recent diagnostic devices, etc., to make more prompt choices regarding the priority of the patients. Predictive workload models derived from continuous monitoring data enable institutions to anticipate surges in demand during pandemics or emergencies. This foresight allows for proactive planning, balanced staffing, and equitable distribution of care, ensuring that critically ill patients have timely and equitable access to intensive care resources.

However, in the **Intensive Care Unit,** the following resources as shown in Figure 4 are typically prioritized for optimization:

1. **Human Resources**
   a) Critical care physicians, nurses, and specialists – optimal staffing ratios are required to ensure safe patient monitoring and interventions in the ICU.
   b) Respiratory therapists and technicians – particularly crucial for ventilator management.
   c) Support staff – including pharmacists, physiotherapists, and dieticians to deliver multidisciplinary care.
   d) Scheduling and workload distribution – balancing staff fatigue with patient care demands.
2. **Bed Capacity and Space Utilization**
   a) ICU beds – ensuring availability during surges (e.g., pandemics).
   b) Isolation rooms – efficient use for infectious disease control.
   c) Allocation algorithms – prioritizing critically ill patients when demand exceeds supply.

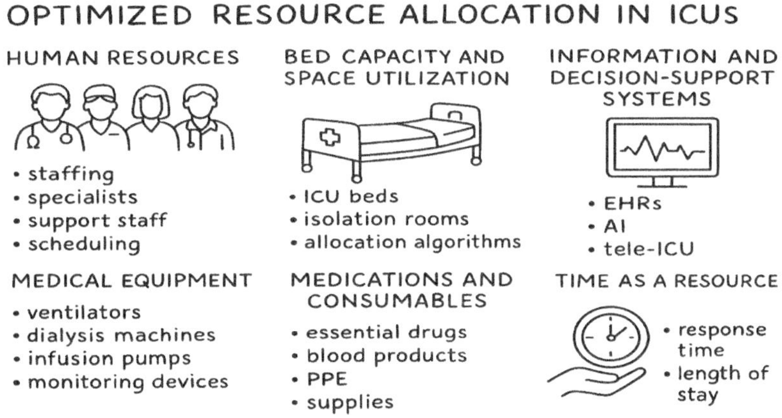

Figure 4: Optimized resource allocation in ICUs.

3. **Medical Equipment**
   a) Ventilators – optimized distribution and monitoring of usage.
   b) Dialysis machines and infusion pumps – assigned according to the acuity.
   c) Higher level monitoring- ECG, oxygenation monitors, etc., to make decisions on time.
   d) Access to imaging equipment – e.g., bedside diagnostics with portable X-rays or ultrasound.
4. **Medications and Consumables**
   a) Essential drugs – antibiotics, sedatives, vasopressors, analgesics.
   b) Blood products – important to use rationally to prevent shortages.
   c) Personal protective equipment (PPE) – rational distribution of personal protective equipment in case of an infectious outbreak.
   d) Medical materials that can be disposed of – syringes, catheters, suction, and so on.
5. **Information and Decision-Support Systems**
   a) Electronic Health Records (EHRs) – to prevent redundancy and improve care.
   b) AI-driven decision-support systems – for predictive analytics (e.g., deterioration alerts, length-of-stay predictions).
   c) Tele-ICU systems – to spread experience where there is a shortage of specialists.
6. **Time as a Resource**
   a) Reaction time to crisis situations - workflow streamlining to reduce the time spent.
   b) ICU stay – timing of discharge: the patient would be released early, opening up more beds to receive new ones.

Automated monitors and DSSs enhance intensive care beyond assisting patient safety or lessening the prevalence of human error or higher levels of predictiveness to present anticipatory care. They optimize the utilization of resources and expand resource capacity during high demand, which is aligned with the vision of precise healthcare. This makes the ICU an inclusive and knowledgeable facility applicable to the needs of patients and the whole healthcare system.

Intensive care units (ICUs) generate enormous amounts of patient data, presenting both opportunities and challenges for automated monitoring and decision support. However, these systems face numerous limitations and risks. Key obstacles include inconsistent data quality and lack of interoperability, as well as ethical and technical concerns that can undermine system reliability and trust.

7. Challenges and Limitations

1. **Data quality, interoperability, and standardization issues:** ICU data come from diverse sources (monitors, ventilators, labs, EHRs) with different formats and structures (Cecconi et al. 2025). Variations in device outputs, charting practices, and local policies mean that data often lack uniformity and completeness (Cecconi et al. 2025, Lijović and Elbers 2025). For example, oxygenation and ventilator data may use proprietary encodings, making automated aggregation difficult. To address this, common data models and standards (e.g., OMOP, LOINC, SNOMED CT, and HL7 FHIR) are increasingly adopted to harmonize data (Cecconi et al. 2025, Lijović and Elbers 2025). Nevertheless, broad adoption is far from universal, so many ICU datasets suffer from missing fields, inconsistent units, and outliers. These quality issues undermine predictive algorithms: as one expert notes, "garbage in, garbage out" means that flawed or incomplete data can lead to faulty outputs regardless of analytic sophistication (Lijović and Elbers 2025). Until interoperability is improved, hospitals must carefully validate and curate data before feeding it to decision-support tools (Lijović and Elbers 2025).
2. **Risk of overreliance on automated systems:** Clinical staff may be tempted to *trust* automated alerts and recommendations without question. However, AI and automation are far from infallible. Studies of AI-enabled clinical decision support warn that excessive dependence can erode clinicians' critical judgment (Hassanein et al. 2025). Overreliance may lead providers to accept an alert even if it contradicts subtle clinical cues (so-called "automation bias"), or to ignore non-algorithmic signs of deterioration. For example, nurses worry that trusting a monitoring algorithm too much could "erode critical thinking and clinical expertise" (Hassanein et al. 2025). This can degrade patient care by missing nuanced needs that only human experience can catch. Hence, experts emphasize that AI systems should be *assistive* aids, not replacements, ensuring human oversight (Hassanein et al. 2025, Gutierrez 2020). Continuous training is needed so that ICU staff maintain their own assessment skills and view AI as a "supportive adjunct" rather than an oracle_(Gutierrez 2020,_Hassanein et al. 2025).
3. **Ethical considerations: Patient privacy, informed consent, and algorithmic transparency:** The deployment of automated monitoring raises important ethical questions. Patient data used in AI models must be protected under strict privacy regulations. For instance, breaches or unauthorized data use in the ICU would be catastrophic—akin to losing oxygen in an ICU setting (Cecconi et al. 2025). Therefore, robust security and encryption are essential, and patients' privacy rights must be honored in any data pipeline (Cecconi et al. 2025). Consent is another challenge: in emergency ICU settings, it may be unclear how or when to inform patients or proxies about AI-assisted decisions (Cecconi et al. 2025). Equally critical is **algorithmic transparency**. Clinicians express wariness

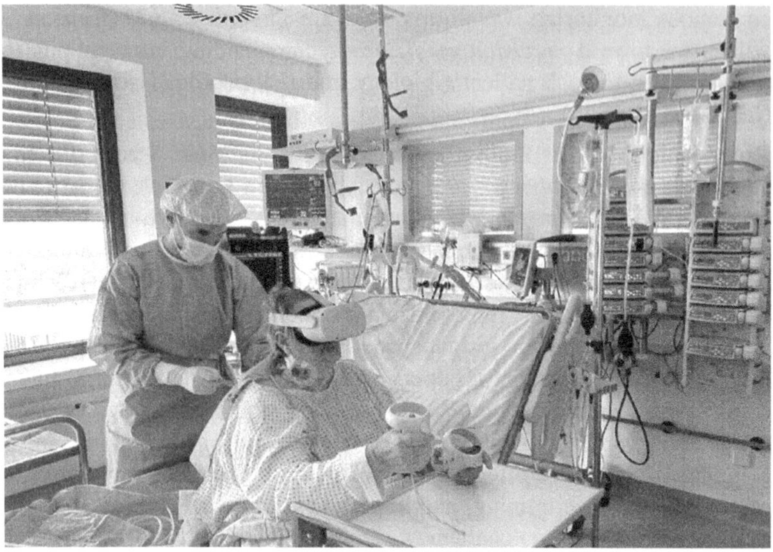

Figure 5: An example of AR/VR applications in intensive care (Kanschik et al. 2023).

example, robotic systems could deliver medications or assist with hygiene, while algorithms dynamically titrate ventilators and IV drips based on feedback loops. Early prototypes ("semi-autonomous ICUs") are already being piloted: one study describes robots that bring supplies to bedsides on command, and AI systems that can self-configure and recover from failures (Gutierrez 2020). A visionary 2020 article predicted that developing continuous AI-based monitors would pave the way for an ICU where "intelligent machines provide most of the care delivered today by humans" (Gutierrez 2020). Such a future would reduce human workload and potentially improve consistency of care. However, realization of fully autonomous ICUs will require solving current challenges (reliability, ethics, integration). For now, the trend is toward *smart ICUs* that layer AI and robotics on top of human-centered care, with machines taking over routine monitoring and alerting, but final decisions remaining in clinicians' hands (Gutierrez 2020).

9. Conclusion

The integration of automated patient monitoring and decision support systems (DSS) represents a transformative shift in intensive care delivery. Traditional ICU monitoring, while essential, often suffers from fragmentation, data overload, and alarm fatigue, which can compromise both patient safety and clinician efficiency. Automated systems, powered by biosensors, IoT devices, artificial intelligence, and real-time analytics, address these gaps by providing continuous, accurate, and predictive insights into patient health. Decision support tools further enhance care by offering evidence-based recommendations, early warnings, and personalized treatment pathways, ultimately improving survival rates and reducing human error. Despite challenges such as interoperability issues, ethical concerns, and risks of overreliance, these technologies promise to optimize resource allocation, support proactive interventions, and enhance staff well-being. Looking ahead, innovations such as digital twins, AR/VR integration, and precision medicine will further redefine ICU care, making it more adaptive, predictive, and patient-centered. Thus, the future of critical care lies in a balanced synergy between human expertise and intelligent systems, ensuring safer, more efficient, and more resilient healthcare delivery.

References

Aakula, R. (2025). Real-Time AI Dashboards for ICU Monitoring and Alerting. *European Journal of Computer Science and Information Technology*, 13(12): 15–23.

Adeniyi, E. A., Ogundokun, R. O., and Awotunde, J. B. (2021). IoMT-based wearable body sensors network healthcare monitoring system. In IoT in healthcare and ambient assisted living (pp. 103–121). Singapore: Springer Singapore.

Adler-Milstein, J., Worzala, C., and Dixon, B. E. (2023). Future directions for health information exchange. In *Health Information Exchange* (pp. 447–468). Academic Press.

Ahmed, F., Hossain, M. S., Islam, R. U., and Andersson, K. (2021). An evolutionary belief rule-based clinical decision support system to predict COVID-19 severity under uncertainty. *Applied Sciences*, 11(13): 5810.

AL-Dossary, R. N. (2023). Nurses' Perceptions of the Clinical Decision Support System Effect on Patient Safety. *Safety*, 9(4): 86.

Archana, S. K. S. (2024). Artificial Intelligence in Critical Care: Enhancing Decision-Making and Patient Outcomes. *European Journal of Cardiovascular Medicine*, 14: 497–505.

Armando, L. G., Miglio, G., de Cosmo, P., and Cena, C. (2023). Clinical decision support systems to improve drug prescription and therapy optimisation in clinical practice: a scoping review. *BMJ Health and Care Informatics*, 30(1)– e100683.

Awotunde, J. B., Jimoh, R. G., Ogundokun, R. O., Misra, S., and Abikoye, O. C. (2022). Big data analytics of iot-based cloud system framework: Smart healthcare monitoring systems. In *Artificial intelligence for cloud and edge computing* (pp. 181–208). Cham: Springer International Publishing.

Ball, K. (2021). Electronic monitoring and surveillance in the workplace. *European Commission Joint Research Centre*.

Barrit, S. (2025). *Machine Learning and Artificial Intelligence in Intensive Care Medicine: Critical Recalibrations from Rule-Based Systems to Frontier Models. Journal of Clinical Medicine,* 14(12): 4026. https://doi.org/10.3390/jcm14124026

Bignami, E., Lanza, R., Cussigh, G., and Bellini, V. (2023). New technologies in anesthesia and intensive care: take your ticket for the future. *Journal of Anesthesia, Analgesia and Critical Care*, 3(1). https://doi.org/10.1186/s44158-023-00098-3

Brown, T. (2019). Automation in Intensive Care Units. Medical Engineering and Informatics.

Callcut, R. A., Xu, Y., Moorman, J. R., Tsai, C., Villaroman, A., Robles, A. J. et al. (2021). External validation of a novel signature of illness in continuous cardiorespiratory monitoring to detect early respiratory deterioration of ICU patients. *Physiological Measurement*, 42(9). https://doi.org/10.1088/1361-6579/ac2264

Ch. Gayathri Sowmya Sree, Lakshmi Lavanya, G., Srinivasa Reddy, K., Sai Kiran, K., and Sirisha, M. (2022). IoT-Based ICU patient monitoring system. *International Journal of Engineering Technology and Management Sciences*, 106–109. https://doi.org/10.46647/ijetms.2022.v06si01.021

Cecconi, M., Greco, M., Shickel, B., Angus, D. C., Bailey, H., Bignami, E. et al. (2025). *Implementing artificial intelligence in critical care medicine: a consensus of 22. Critical Care,* 29, Article 290. https://doi.org/10.1186/s13054-025-05532-2

Chen, Z., Liang, N., Zhang, H., Li, H., Yang, Y., Zong, X. et al. (2023). Harnessing the power of clinical decision support systems: challenges and opportunities. *Open Heart*, 10(2): e002432.

Coffman, L. (2018). Patient wearable technology in the ICU. *Nursing Critical Care*, 13(5): 41–42. https://doi.org/10.1097/01.CCN.0000534929.31666.37

Collaguazo, A., Estrada, R., and Arreaga, N. X. (2022). IoT-based Health Monitoring System for Intensive Care Units. In *WINSYS* (pp. 101–106).

Cuvelier, E., Robert, L., Musy, E., Rousseliere, C., Marcilly, R., Gautier, S. et al. (2021). The clinical pharmacist's role in enhancing the relevance of a clinical decision support system. *International Journal of Medical Informatics, 155*, 104568.

De Filippis, R., and Al Foysal, A. (2025). AI-Driven Early Warning and Risk Management System for Delirium in ICU Patients. *Open Access Library Journal*, 12(1): 1–17.

Defense, D., Yin, Y., Members, D. C., Chou, C., Griffin, J., and Xie, W. (2022). *Data-Driven Modeling and Learning Approach to Solving ICU Risk Prediction and Survival Analysis.*

Dewan, M., Vidrine, R., Zackoff, M., Paff, Z., Seger, B., Pfeiffer, S. et al. (2020). Design, implementation, and validation of a pediatric ICU sepsis prediction tool as clinical decision support. *Applied Clinical Informatics*, 11(02): 218–225.

Diao, J. A., Marwaha, J. S., and Kvedar, J. C. (2022). Video-based physiologic monitoring: promising applications for the ICU and beyond. *Npj Digital Medicine*, 5(1). https://doi.org/10.1038/s41746-022-00575-z

Elhaddad, M., and Hamam, S. (2024). AI-driven clinical decision support systems: an ongoing pursuit of potential. *Cureus*, 16(4).

Goyal, S., and Fiorini, L. (2025). Practical Implementation and Integration of AI in Mental Healthcare. In *Adversarial Deep Generative Techniques for Early Diagnosis of Neurological Conditions and Mental Health Practises: Theoretical Insights with Practical Applications* (pp. 373–415). Cham: Springer Nature Switzerland.

Grechuta, K., Shokouh, P., Alhussein, A., Müller-Wieland, D., Meyerhoff, J., Gilbert, J., and Rolland, C. (2024). Benefits of clinical decision support systems for the management of noncommunicable chronic diseases: targeted literature review. *Interactive Journal of Medical Research*, 13(1): e58036.

Gutierrez, G. (2020). *Artificial Intelligence in the Intensive Care Unit. Critical Care,* 24(1), Article 101. https://doi.org/10.1186/s13054-020-2785-y

Halpern, G. A., Nemet, M., Gowda, D. M., Kilickaya, O., and Lal, A. (2024). Advances and utility of digital twins in critical care and acute care medicine: a narrative review. *Journal of Yeungnam Medical Science*, 42.

Halpern, G. A., Nemet, M., Gowda, D. M., Kilickaya, O., and Lal, A. (2025). *Advances and utility of digital twins in critical care and acute care medicine: A narrative review. Journal of Yeungnam Medical Science,* 42, Article 9. https://doi.org/10.12701/jyms.2024.01053

Hassanein, S., El Arab, R. A., Abdrbo, A., Abu-Mahfouz, M. S., Gaballah, M. F., Seweid, M. M. et al. (2025). Artificial intelligence in nursing: An integrative review of clinical and operational impacts. *Frontiers in Digital Health*, 7. https://doi.org/10.3389/fdgth.2025.1552372

Henry, B., Merz, M., Hoang, H., Abdulkarim, G., Wosik, J., and Schoettker, P. (2024). Cuffless Blood Pressure in clinical practice: challenges, opportunities and current limits. *Blood Pressure*, 33(1): 2304190.

Hosain, M. N., Kwak, Y. S., Lee, J., Choi, H., Park, J., and Kim, J. (2024). IoT-enabled biosensors for real-time monitoring and early detection of chronic diseases. *Physical Activity and Nutrition*, 28(4): 60.

Huo, J., Wung, S., Roveda, J., and Li, A. (2023). Reducing false alarms in intensive care units: A scoping review. *Exploratory Research and Hypothesis in Medicine*, 8(1): 57–64.

Idelchik, M., and Kogan, S. (2012). GE's open collaboration model. *Research Technology Management*, 55(4): 28–31. https://doi.org/10.5437/08956308X5504101

Jimenez-Pastor, A., Cerdá-Alberich, L., Kosvyra, A., Nan, Y., Xing, X., Li, R. et al. (2025). Data Harmonization and Challenges Toward the Generation of Repositories: Sharing Practices and Approaches. In *Trustworthy AI in Cancer Imaging Research* (pp. 121–142). Cham: Springer Nature Switzerland.

Jung, Y. H., Kim, J. U., Lee, J. S., Shin, J. H., Jung, W., Ok, J. et al. (2020). Injectable biomedical devices for sensing and stimulating internal body organs. *Advanced Materials*, 32(16): 1907478.

Kanschik, D., Bruno, R. R., Wolff, G., Kelm, M., and Jung, C. (2023). *Virtual and augmented reality in intensive care medicine: A systematic review. Annals of Intensive Care,* 13, Article 81. https://doi.org/10.1186/s13613-023-01176-z

Kaur, C. (2020). The Cloud Computing and Internet of Things (IoT). *International Journal of Scientific Research in Science, Engineering and Technology*, 19–22. https://doi.org/10.32628/ijsrset196657

Khan, Z. (2025). Real-time AI systems for ICU risk monitoring. *Int. J. Multidiscip. Res.*, *7*(2): 1–10.

Khope, S. R., and Elias, S. (2022). Critical Correlation of Predictors for an Efficient Risk Prediction Framework of ICU Patient Using Correlation and Transformation of MIMIC-III Dataset. *Data Science and Engineering*, 7(1): 71–86. https://doi.org/10.1007/s41019-022-00176-6

Lee, A., and Kumar, R. (2021). Decision Support in Critical Care. International Journal of Healthcare Informatics.

Li, L., Ackermann, K., Baker, J., and Westbrook, J. (2020). Use and evaluation of computerized clinical decision support systems for early detection of sepsis in hospitals: protocol for a scoping review. *JMIR Research Protocols*, 9(11): e24899.

Lijović, L., and Elbers, P. (2025). *Leveraging the power of routinely collected ICU data. Intensive Care Medicine*, 51: 163–166. https://doi.org/10.1007/s00134-024-07745-5

Lin, T. H., Chung, H. Y., Jian, M. J., Chang, C. K., Lin, H. H., Yen, C. T. et al. (2025). AI-driven innovations for early sepsis detection by combining predictive accuracy with blood count analysis in an emergency setting: Retrospective study. *Journal of Medical Internet Research*, 27: e56155.

Liu, W., Tao, G., Zhang, Y., Xiao, W., Zhang, J., Liu, Y. et al. (2022). A simple weaning model based on interpretable machine learning algorithm for patients with sepsis: A research of MIMIC-IV and eICU databases. *Frontiers in Medicine*, 8. https://doi.org/10.3389/fmed.2021.814566

Lu, S., Ranjan, R., and Strazdins, P. (2013). Reporting an experience on design and implementation of e-Health systems on Azure cloud. *Lecture Notes in Computer Science (Including Subseries Lecture Notes in Artificial Intelligence and Lecture Notes in Bioinformatics)*, 8223 LNCS, 145–159. https://doi.org/10.1007/978-3-642-41428-2_12

Luo, M.-H., Huang, D.-L., Luo, J.-C., Su, Y., Li, J.-K., Tu, G.-W. et al. (2022). Data science in the intensive care unit. *World Journal of Critical Care Medicine*, 11(5): 311–316. https://doi.org/10.5492/wjccm.v11.i5.311

Murali, M., Ni, M., Karbing, D. S., Rees, S. E., Komorowski, M., Marshall, D. et al. (2024). Clinical practice, decision-making, and use of clinical decision support systems in invasive mechanical ventilation: a narrative review. *British Journal of Anaesthesia*, 133(1): 164–177.

Nahar, L., Zafar, S. S., and Rafiq, F. B. (2020). IOT Based ICU patient health monitoring system. *11th Annual IEEE Information Technology, Electronics and Mobile Communication Conference, IEMCON 2020*, 407–413. https://doi.org/10.1109/IEMCON51383.2020.9284900

Nishat, A. (2024). AI-powered decision support and predictive analytics in personalized medicine. *Journal of Computational Innovation*, 4(1).

Nouis, S. C. E., Uren, V., and Jariwala, S. (2025). Evaluating accountability, transparency, and bias in AI-assisted healthcare decision-making: A qualitative study of healthcare professionals' perspectives in the UK. *BMC Medical Ethics*, 26, Article 89. https://doi.org/10.1186/s12910-025-01243-z

Olanrewaju, Z. O., Faisal, M., and Randell, R. (2024). A scoping review of the role of clinical decision support systems in intensive care units during the COVID-19 pandemic. *Journal of Applied Interdisciplinary Research*, (Special Issue), 21–34.

Olujide, A. P., Babatunde, A. I., and Babatunde, A. J. (2025). *AI in Wearable Imaging Devices*. 159–194. https://doi.org/10.4018/979-8-3373-0903-3.ch006

Ongenae, F., Gadeyne, B., De Backere, F., De Bus, L., Depuydt, P., Decruyenaere, J. et al. (2015). Digital Dr. House: A semantic alerting platform for the ICU. *CEUR Workshop Proceedings*, 1546: 201–202.

Papadopoulos, P., Soflano, M., Chaudy, Y., Adejo, W., and Connolly, T. M. (2022). A systematic review of technologies and standards used in the development of rule-based clinical decision support systems. *Health and Technology*, 12(4): 713–727.

Poncette, A. S., Mosch, L., Spies, C., Schmieding, M., Schiefenhövel, F., Krampe, H. et al. (2020). Improvements in patient monitoring in the intensive care unit: survey study. *Journal of Medical Internet Research*, 22(6): e19091.

Prasad, N., Cheng, L. F., Chivers, C., Draugelis, M., and Engelhardt, B. E. (2017). A reinforcement learning approach to weaning of mechanical ventilation in intensive care units. *arXiv preprint arXiv:1704.06300*.

Preston, R. (2022). Stifling innovation: how global data protection regulation trends inhibit the growth of healthcare research and start-ups. *Emory Int'l L. Rev.*, 37: 135.

Raju, K., Lakshmi, N. D., Komal, P. D., Hari, B. S., and Devi, S. H. (2023). IoT based ICU patient monitoring system. *J. Eng. Sci.*, 14(04).

personalized treatment protocols, ongoing physiological monitoring, and accurate remote diagnoses. With care being taken to patients' homes and daily lives outside the conventional hospital or clinic environment, the implementation of IoMT solutions is revolutionizing the limits of healthcare provision (Chen et al. 2025).

The wide range of IoMT applications varies from ubiquitous consumer-level devices such as wearable fitness monitors, smartwatches, and intelligent scales to very specialized and sophisticated clinical devices such as Continuous Glucose Monitoring (CGMs), intelligent inhalers, wirelessly networked surgical robots, and sophisticated hospital asset management systems. The foundational promise of IoMT lies in its capacity to seamlessly bridge the existing informational and operational gaps between patients, dedicated healthcare providers, and cutting-edge technology. This integration is designed to foster a synergistic environment that demonstrably improves patient outcomes, significantly enhances the efficiency of healthcare delivery, and ultimately contributes to a reduction in overall healthcare expenditures.

Crucially, IoMT facilitates the widespread adoption of telemedicine and telehealth services, empowering medical professionals to effectively monitor and manage patients from geographically dispersed locations. Furthermore, it significantly augments preventive care strategies by enabling the early detection and identification of potential health issues before they escalate into critical conditions. However, the successful and sustainable implementation of IoMT is intrinsically dependent upon the meticulous development and deployment of a robust, resilient, and inherently secure infrastructure (El-Saleh et al. 2025). In addition to guaranteeing the availability, confidentiality, and integrity of the enormous amount of extremely sensitive health data produced by these devices, this infrastructure must be able to manage it effectively. IoMT and associated technologies are depicted in Figure 1.

The fundamental elements that constitute the IoMT architecture shall be examined thoroughly in this chapter, and each element will be analyzed to determine its purpose and importance. It shall then address the complex issues and major factors that bear an integral role in the seamless and effective integration of IoMT solutions in the current healthcare system. Fog computing creates a hierarchy of multiple layers

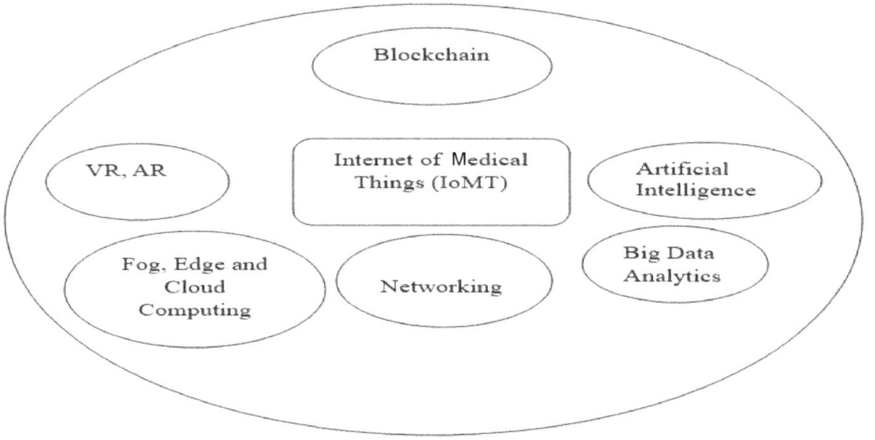

Figure 1: Artificial Intelligence in Cybersecurity.

between hardware and the Cloud server, which reduces data storage across cloud servers and further reduces bandwidth along with response time in cloud computing. Fog computing enhances data security by keeping data locally on the edge. Fog and edge computing emerged as robust and secure paradigms (Dwivedi et al. 2022).

2. Evolution of IoMT

The term Internet of Medical Things (IoMT) explores the interconnected network of healthcare technologies, which include sensors, medical devices, and wearables. IoMT devices can collect a range of information on vital signs, patient behavior, and various environmental factors. This information can be further analyzed in real time to obtain valuable insights about the health status of a patient. IoMT has the capability to enhance the way healthcare is delivered. The origins of IoMT are in simple, frequently stand-alone medical devices with little or no connectivity except within their internal environments.

The wired telematics systems used within managed clinical settings, mostly hospitals, were the main feature of this era. Equipment like bedside cardiac monitors and pulse oximeters would gather vital signs and transmit them to a central monitoring station in the same building over dedicated lines. There was no interoperability that had been realized; proprietary data transport existed, and it was local. Remote access or universal data exchange was practically unheard of, and information flow was mostly unidirectional. In spite of being greatly constrained by physics and isolated data silos, this period set the rudimentary expertise of ongoing patient surveillance (Bin Heyat et al. 2025).

2.1 The Evolution of Wireless Communication

Wireless communication technologies like Bluetooth (IEEE 802.15.1) and Wi-Fi (IEEE 802.11) evolved rapidly and were offered widely. Medical devices also experienced a major transformation due to this technological growth, which increased their portability and versatility and allowed them to break the limitation of wired connections. The first generation of wearable health sensors and basic home-based monitoring devices began to emerge during this period. Patients could even record health data on their own at home due to innovations such as early smart glucometers, Bluetooth-enabled digital blood pressure monitors, and simple activity trackers. In certain conditions, these devices may even send data wirelessly to central servers or personal computers, laying the foundation for modern connected healthcare systems. It was still difficult to incorporate this data into a larger healthcare system, though. Data was often still siloed, requiring manual upload or proprietary software for basic visualization. Interoperability between different manufacturers' devices was a significant hurdle, and data security, while recognized, was not as rigorously addressed as it is today. Despite these limitations, this phase was instrumental in demonstrating the feasibility and potential benefits of remote patient monitoring, extending the reach of care beyond the physical confines of a clinic (Sallabi et al. 2025).

2.2 Convergence with Cloud, Big Data, and Mobile Technologies

The most transformative phase of IoMT evolution is defined by the convergence of several advanced technological paradigms: cloud computing, big data analytics, widespread smartphone, and Artificial Intelligence (AI), as shown in Figure 2.

a. **Cloud Computing:** The advent of scalable and secure cloud platforms (e.g., AWS, Azure, Google Cloud) revolutionized data storage and processing capabilities. Healthcare organizations could now offload the immense volume of health data generated by IoMT devices to centralized, highly available, and resilient cloud infrastructures. This not only provided massive storage capacity but also enabled ubiquitous access to data for authorized users from any location with internet connectivity.

b. **Big Data Analytics:** With large datasets stored in the cloud, the emphasis switched from merely gathering data to deriving insights that could be put to use. Big data analytics methods, such as statistical modeling and pattern identification, have become essential for deciphering intricate physiological data, spotting patterns, and anticipating possible declines in health.

c. **Mobile Technologies:** A key entry point for IoMT was the pervasiveness of smartphones and tablets. Patients now primarily use mobile applications to connect with their care teams, access their health data, and engage with their devices. For healthcare providers, mobile devices enabled real-time access to patient data, facilitating remote consultations and prompt interventions.

d. **Artificial Intelligence (AI) and Machine Learning (ML):** The integration of AI and ML algorithms has been a game-changer for IoMT. These advanced computational techniques are now leveraged to analyze complex datasets, identify subtle patterns indicative of disease, provide predictive insights (e.g., predicting diabetic ketoacidosis), assist in diagnostics, and even personalize

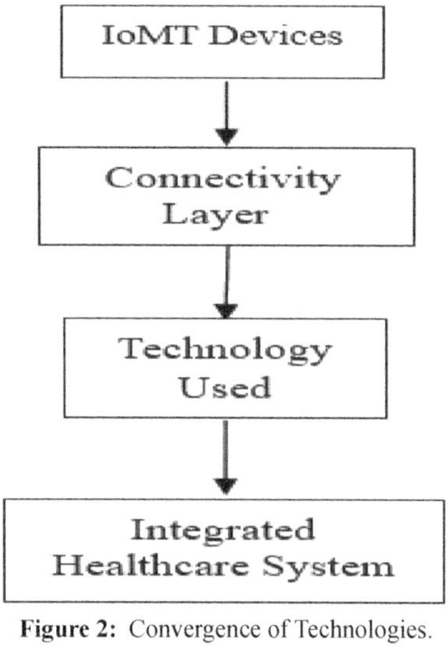

Figure 2: Convergence of Technologies.

treatment recommendations. AI-powered algorithms can process data from multiple sources (wearables, EHRs, imaging) to create a more holistic view of a patient's health.

e. **Advanced Connectivity (4G/5G):** The continuous improvement in cellular network capabilities, particularly the rollout of 4G LTE and the ongoing deployment of 5G, has been critical. 5G, with its characteristics of ultra-low latency, massive connectivity, and significantly higher bandwidth, promises to unlock new frontiers for IoMT. It enables real-time applications such as high-definition remote surgical assistance, real-time feedback for physical therapy, and the seamless operation of a dense network of sensors in smart hospitals.

This current phase represents a shift from simple data transfer to an intelligent, automated, and highly integrated healthcare delivery system. In order to achieve the goal of truly "smart" healthcare, the emphasis is now on developing a seamless ecosystem where platforms, devices, and human expertise collaborate to provide proactive, individualized, and effective care.

3. Infrastructural Components of IoMT

The backbone of the Internet of Medical Things (IoMT) ecosystem is a secure and robust infrastructure. Each of the many interdependent and interrelated parts that make up this infrastructure is essential to the gathering, sharing, processing, analyzing, and applying of health data. Designing, implementing, and overseeing successful IoMT solutions requires an understanding of these elements.

3.1 Wearable and Implantable Devices

Smart sensor-based technologies present the core foundation of IoT-enabled monitoring systems for accurate and real-time data collection across diverse environments. These sensors are designed to detect environmental parameters including temperature, humidity, and air pollutants (soil moisture and radiation levels). With the growing emphasis on sustainability and precision, the development of advanced sensors using micro-electromechanical systems (MEMS), nanomaterials, and biodegradable compounds has significantly improved their sensitivity, durability, and environmental compatibility (Rao et al. 2025).

3.1.1 Wearable Sensors

These are non-invasive devices worn on the body. Examples include:

a. **Smartwatches and Fitness Trackers:** These devices are used to monitor heart rate, sleep patterns, and in some advanced models, ECG and blood oxygen saturation.

b. **Continuous Glucose Monitors (CGMs):** These devices are used to monitor the glucose levels and provide real time data for diabetes management.

4.3 Artificial Intelligence (AI) and Machine Learning (ML)

a. **Pattern Recognition:** ML algorithms excel at identifying subtle, complex patterns in vast datasets that might be imperceptible to humans. This is essential for using sensor data to detect diseases early (Ullah et al. 2024).

b. **Anomaly Detection:** Identifying deviations from a patient's normal physiological baseline, triggering alerts for potential issues.

c. **Diagnostic Support:** AI algorithms can assist clinicians by analyzing medical images (e.g., X-rays, MRIs), ECGs, or voice patterns to identify abnormalities.

d. **Personalized Interventions:** ML models can tailor treatment plans, medication dosages, or lifestyle recommendations based on individual patient data, genetics, and response to therapies.

e. **Natural Language Processing (NLP):** Used to gather useful information from patient narratives and unstructured clinical notes, then combine it with sensor data to create a more comprehensive medical profile.

5. Integration Challenges and Considerations

IoMT is full of promise, but there are numerous hurdles to be crossed before it becomes widely accepted and implemented successfully, particularly regarding aggregating multiple platforms, devices, and streams of data into a unified healthcare system. These challenges require proper planning, robust technology, and integration of all segments of the healthcare industry.

5.1 Interoperability

The complete interoperability in IoMT integration is the most challenging task, as healthcare information derives from various sources.

a. **Syntactic Interoperability:** It is the ability of systems to exchange information in any structure. Most of the legacy clinical devices use proprietary protocols, due to which data extraction and integration is challenging.

b. **Organizational Interoperability:** It deals with legal policies and organizational aspects to enable secure and useful sharing of information among different organizations. This involves data governance frameworks, consent management, and agreement on workflows.

c. **Technical Integration:** The practical challenge of connecting devices and systems that may have different operating systems, hardware specifications, and software APIs. This often requires middleware, gateways, and custom development to translate and route data.

Lack of interoperability leads to data silos, inhibits comprehensive patient views, increases manual data entry errors, and prevents the full realization of IoMT's potential for analytics and personalized care.

5.2 Data Security and Privacy

Healthcare data is among the most sensitive and valuable forms of personal information, making it a prime target for cyberattacks. Ensuring its security and privacy is paramount (Ozcelik et al. 2025).

a. **Cybersecurity Threats:** IoMT devices, often resource-constrained, can be vulnerable entry points for attackers. Threats include data breaches, ransomware attacks, Denial-of-Service (DoS) attacks, and even physical tampering. A single compromised device could expose vast amounts of patient data or be used to disrupt critical healthcare services.

b. **Encryption:** Data must be encrypted both at rest (when stored on devices or in the cloud) and in transit (as it moves across networks).

c. **Access Control and Authentication:** Access restrictions and robust authentication methods such as multi-factor authentication allow only authorized users and systems to access certain information.

d. **Privacy-by-Design:** Incorporating the concern of privacy in IoMT devices and platforms should be from the beginning, not at a later stage. It further requires the data usage to be more transparent (Sudharson et al. 2025).

e. **Regulatory Compliance:** Compliance with data protection laws, including national or local privacy laws, the European Union's General Data Protection Regulation (GDPR), and the United States' Health Insurance Portability and Accountability Act (HIPAA). Non-compliance can lead to serious legal penalties.

f. **Ethical concerns:** Bias in AI healthcare solutions, data ownership, and secondary use of data for commercial or research purposes are ethical concerns. Patient consent must be taken through clear and informed procedures.

5.3 Compliance with Regulations and Standards

To maintain patient safety and data integrity, the medical and healthcare industries are required to adhere to stringent regulations. It is a tangled web of global and domestic regulations that IoMT devices and systems must go through.

a. **Medical Devices Regulations:** IoMT devices can be deemed medical devices and need approval from agencies such as the European Medicines Agency (EMA) (for CE marking in Europe), the U.S. Food and Drug Administration (FDA), or equivalents worldwide, depending on intended use and risk category. This includes clinical validation, rigorous testing, and compliance with quality management systems (such as ISO 13485).

b. **Software as a Medical Device (SaMD):** Software components of IoMT platforms, even if not physically attached to a device, can be classified as SaMD, requiring their own regulatory clearances.

c. **Cybersecurity Standards:** Regulators are increasingly focusing on the cybersecurity posture of medical devices. The FDA, for instance, provides guidance on premarket and postmarket cybersecurity management for medical devices.

d. **Data Standards:** Adherence to established healthcare data standards (e.g., DICOM for imaging, LOINC for lab tests, SNOMED CT for clinical terms) is crucial for facilitating interoperability and meaningful data exchange.

e. **Interoperability Frameworks:** Efforts like the Argonaut Project (based on FHIR) aim to accelerate the adoption of common standards for healthcare data exchange, which are vital for IoMT integration.

Navigating these regulatory complexities can be time-consuming and expensive, acting as a barrier to innovation and market entry for new IoMT solutions.

5.4 Scalability and Reliability

As IoMT deployments grow, the underlying infrastructure must be capable of handling increasing demands without compromising performance or stability.

a. **Data Volume and Velocity:** A single patient's continuous monitoring can generate gigabytes of data daily. Scaling to thousands or millions of patients requires robust data pipelines, storage solutions, and processing capabilities.

b. **System Uptime and Availability:** In critical healthcare scenarios, IoMT systems must be continuously available. Downtime can have severe consequences for patient safety. This necessitates redundant systems, failover mechanisms, and robust disaster recovery plans.

c. **Device Management:** Managing a vast fleet of diverse IoMT devices (provisioning, updates, troubleshooting, battery management) poses a significant logistical and technical challenge. Over-the-air (OTA) updates are crucial but must be implemented securely.

d. **Network Resilience:** Ensuring that network connectivity is always available, even in adverse conditions. This involves designing networks with redundancy and utilizing technologies that can perform in challenging environments (Gilbert et al. 2024).

5.5 User Acceptance and Training

The most sophisticated IoMT system is ineffective if healthcare professionals and patients are unwilling or unable to use it effectively.

a. **Clinical Workflow Integration:** IoMT solutions must seamlessly integrate into existing clinical workflows rather than creating additional burdens. This requires intuitive interfaces, minimal data entry, and clear benefits for clinicians.

b. **Digital Literacy:** Both patients and healthcare providers may have varying levels of digital literacy. To guarantee comfort and competence with IoMT devices and platforms, training and continuing support are essential.

c. **Confidence and Trust:** Patients need to believe in the reliability of technology and the safety of their information. Physicians need to believe in the accuracy and usefulness of information generated by IoMT.

d. **Patient Engagement:** Ongoing patient participation is necessary for remote monitoring programs. To motivate patients to adhere to monitoring processes, this requires simple devices to use, clear instructions, and benefits that are perceived.

e. Establishing an integrated and successful IoMT ecosystem requires a multidisciplinary approach involving collaboration among technology developers, physicians, lawmakers, and patients themselves.

6. Applications and Impact of IoMT

IoMT's long list of uses in different areas of healthcare realizes its transformative potential. By improving efficiency, enhancing patient care, and broadening the reach of healthcare services, these applications are transforming the way medical care is being provided.

6.1 Management of Chronic Diseases

One of the areas where IoMT is having the greatest impact is the management of chronic diseases, which are the leading source of patient burden and healthcare costs.

a. **Diabetes Management:** Healthcare providers and patients can monitor trends, detect patterns, and quickly change insulin doses, diet, or activity levels with the assistance of Continuous Glucose Monitoring (CGMs), which provide immediate glucose values. IoMT offers a closed-loop mechanism for extremely personalized diabetes treatment when used with smart insulin pens or pumps.

b. **Cardiovascular Disease (CVD) Management:** IoMT devices such as smart blood pressure cuffs, ECG patches, and wearable heart rate monitors allow for continuous monitoring of vital signs, enabling the detection of arrhythmias, hypertension, or indicators of heart failure exacerbation early on. Clinicians can initiate preventive actions to avert significant cardiac events using this information.

c. **Respiratory Disease:** For asthma and COPD conditions, smart inhalers monitor technique and drug adherence. Wearable sensors can be used to monitor the breathing rate and oxygen saturations to provide alerts in situations when serious medical attention is required.

d. **Remote Monitoring of Multiple Sclerosis (MS) and Parkinson's Disease:** Various advanced wearable devices, like smartwatches or sensors, can measure activity levels. The collection of this data can provide clear information about disease progression and medication response. This helps to fine-tune the treatment plan for the best possible outcome.

6.2 Assisted Living and Elderly Care

In addition to giving peace of mind, IoMT has a key role to play in allowing older adults to remain safe and independent in their preferred location.

a. **Fall Detection Systems:** Prior incidence sensing and automated alert systems can reduce response time and decrease the risk of complications. By sensing falls and automatically alerting caretakers or emergency services, wearable

IoMT holds an extremely promising future, with ongoing advancements in AI, Blockchain, digital twins, and robotics that will push the limits of medicine. Through the evolution of these technologies, IoMT will enable fully personalized medicine, self-governing therapeutic and diagnostic interventions, and an improved, more accessible, and more egalitarian worldwide healthcare system. The journey to fully integrated IoMT is multifaceted and requires collaboration from patients, clinicians, technologists, and legislators. But the overwhelming benefits—improved patient outcomes, enhanced quality of life, and sustainable healthcare provision—render this revolutionary project a social imperative rather than simply a technological aspiration.

References

Bin Heyat, M. B., Adhikari, D., Akhtar, F., Parveen, S., Zeeshan, H. M., Ullah, H. et al. (2025). Intelligent internet of medical things for depression: current advancements, challenges, and trends. *International Journal of Intelligent Systems*, 2025(1): 6801530.

Chen, Z., Hao, J., Sun, H., Li, M., Zhang, Y., and Qian, Q. (2025). Applications of digital health technologies and artificial intelligence algorithms in COPD: systematic review. *BMC Medical Informatics and Decision Making*, 25(1): 77.

Dwivedi, R., Mehrotra, D., and Chandra, S. (2022). Potential of Internet of Medical Things (IoMT) applications in building a smart healthcare system: A systematic review. *Journal of Oral Biology and Craniofacial Research*, 12(2): 302–318.

El-Saleh, A. A., Sheikh, A. M., Albreem, M. A., and Honnurvali, M. S. (2025). The internet of medical things (IoMT): opportunities and challenges. *Wireless Networks*, 31(1): 327–344.

Gilbert, C., and Gilbert, M. A. (2024). The integration of blockchain technology into database management systems for enhanced security and transparency. *International Research Journal of Advanced Engineering and Science*, 9(4): 316–334.

Hossain, M. D., Rahman, M. H., and Hossan, K. M. R. (2025). Artificial Intelligence in healthcare: Transformative applications, ethical challenges, and future directions in Medical Diagnostics and Personalized Medicine.

Mathkor, D. M., Mathkor, N., Bassfar, Z., Bantun, F., Slama, P., Ahmad, F. et al. (2024). Multirole of the internet of medical things (IoMT) in biomedical systems for managing smart healthcare systems: An overview of current and future innovative trends. *Journal of Infection and Public Health*, 17(4): 559–572.

Mazhar, T., Shah, S. F. A., Inam, S. A., Awotunde, J. B., Saeed, M. M., and Hamam, H. (2024). Analysis of integration of IoMT with blockchain: issues, challenges and solutions. *Discover Internet of Things*, 4(1): 21.

Ozcelik, M. M., Kok, I., and Ozdemir, S. (2025). A Survey on Internet of Medical Things (IoMT): Enabling technologies, security and explainability issues, challenges, and future directions. *Expert Systems*, 42(5): e70010.

Rao, P. V., Khan, S. A., Paul, S., Mitra, S. P., Deb, D., Chaudhuri, A. K. et al (2025). Iot-enabled environmental monitoring systems: trends, challenges, and future directions. *International Journal of Environmental Sciences*, 11(17s), 2025.

Sallabi, F. M., Khater, H. M., Tariq, A., Hayajneh, M., Shuaib, K., and Barka, E. S. (2025). Smart Healthcare Network Management: A Comprehensive Review. *Mathematics*, 13(6): 988.

Sudharson, K., Babu, G., Santhiya, R., and Anita, C. (2025). Enhanced privacy-preserving federated convivial learning for internet of medical things (IoMT) through blockchain-enabled trust Q-learning. *Journal of the National Science Foundation of Sri Lanka*, 52(4).

Ullah, I., Khan, I. U., Ouaissa, M., Ouaissa, M., and El Hajjami, S. (Eds.). (2024). *Future Communication Systems Using Artificial Intelligence, Internet of Things and Data Science*. CRC Press.

Chapter 17

Applications of Healthcare 5.0 in the Context of Industry 5.0

Background, Vision, and Scope

Prince Kumar Yadav,[1] *Mohammad Ashif Iqbal,* [1]
Wasiullah Aman,[1] *Javed Ahmad*[1] and *Keshav Gupta*[2,*]

1. Introduction

The world is on the way to rapid evolution caused by the innovations in the sphere of technology, shifting needs of society and people. Every business sector is reexamining the conventional approaches to how they can serve and deliver services, attempting to seek a middle ground to provide efficiency, sustainability, and personalization. Industry 5.0 is one of such developments, and this new concept has united humans and advanced machines. The healthcare industry is among those that have been influenced by this change the most with the implementation of new technologies such as AI, robotics, Internet of Medical Things (IoMT), and cloud computing, as well as digital twins, transforming the process of care delivery, management, and experience. This paradigm is referred to as Healthcare 5.0, and it goes past automation whereby the patients are central to care with health services being predictive, preventive, personalized, and participatory. The authors in the following subsections elaborate on the principles of Industry 5.0 and healthcare systems development up to Healthcare 5.0 and the vision and ambitions of the new model in addressing modern challenges (Haleem and Javaid 2019).

---

[1] Department of Computer Science & Engineering Sharda School of Computing Science and Engineering Sharda University, Greater Noida, India.
[2] Amity School of Engineering and Technology, Amity University, Mohali, Punjab, India.
Emails: princekumaryadavv@gmail.com; ashifiqbal0786@gmail.com; wasiullahaman432@gmail.com; javed.jmi08@gmail.com
* Corresponding author: keshavgupta101@gmail.com

DOI: 10.1201/9781003783855-21

1.1 Motivation of the Study and Authors' Contribution

The justification behind this paper is that it is a critical thinking of the relationship between Healthcare 5.0 and the human vision of Industry 5.0 and the relationship of the paper to the pressing issues of existing healthcare systems. The traditional systems can no longer be sufficient in the world of the effects of the COVID-19 pandemic, the aging population, and the increased expectations of efficient, personalized, and sustainable healthcare. The need has been desperate to possess smart, connected, and predictive models whereby the patient is at the center of care delivery and at the same time efficient, secure, and accessible in other places.

The authors carry out their contribution to the sphere as they provide a comprehensive review of the history of healthcare between Healthcare 1.0 and Healthcare 5.0 with the help of a well-organized model, tables, and real-life case examples. This chapter discusses the major enabling technologies, such as artificial intelligence (AI), the Internet of Medical Things (IoMT), robotics, blockchain, cloud and edge computing, and digital twins, and explains how these technologies are reshaping the way healthcare services are delivered. Beyond that, it also addresses the main problems of implementing Healthcare 5.0 such as data privacy, interoperability, and ethical concerns, and reflects the views of future research. The chapter emphasizes the importance of multidisciplinary collaboration in order to establish the transition into the sustainable patient-centered Healthcare 5.0 ecosystem to meet the new demands of society.

1.2 Industry 5.0: A Human-Centric Revolution

In the field of production and services, Industry 5.0 is introducing a major shift in the industrial landscape and transforming the human–machine relationship. Unlike in Industry 4.0, where the position of efficiency and mass production was held predominantly by robots, big data analytics, and machine interconnections, in Industry 5.0 the human being is back in the middle of the industrial process. It is not only about maximization of productivity, but the biggest concern is the creation of personalized, resilient, and sustainable systems that enhance the capabilities of people. High-tech equipment such as Artificial Intelligence (AI) and robotics in the new paradigm does not replace humans, but collaborates with them. Workers can be assisted by AI systems in making decisions, as high amounts of data can be analyzed and, moreover, trends and routines forecasted, while human judgment and ethical instincts remain central. Industry 5.0 supports the concept of mass customization; individual products and services are produced based on the requirements of a single individual rather than carrying out a one-size-fits-all production strategy. This change is essential to industries that require human interaction, empathy, and moral decision-making, and one of the most apparent examples may be healthcare (Jagadish et al. 2019).

1.3 Evolution of Healthcare: From Analog to Smart Systems

The factual advancement of the healthcare industry in the past few decades is due to the influence of technological changes and the changes in patient needs. Initially, everything was fully analog as the so-called Healthcare 1.0. The patient records were written manually, diagnosis was determined simply by physician-possessed knowledge, and technology was not involved, and consultations were conducted face-to-face. The transition to Healthcare 2.0 introduced the initial digital technologies to hospitals, the easiest example being electronic medical records (EMRs) and extremely primitive automation of hospital management. This helped easier storage and retrieval of patient data although the system was very fragmented. Further advancement came with Healthcare 3.0 whereby Electronic Health Records (EHRs) and hospital information systems expanded, helping in the organization of patient data in electronic formats to enable easier access and sharing of data within departments. However, these systems still were not real time, interconnected, or predictive. Healthcare 4.0 was a leap into integrating health systems with IoT wearables, telemedicine services, and cloud computing. The advancements made allowed monitoring of patients remotely, initial-level diagnostic AI assistance, and sharing of healthcare data. But Healthcare 4.0 turned out to be highly reactive to symptoms and provided medication as soon as the disease manifested itself, and the data was isolated, with systems and devices not communicating with each other (Schiavoni et al. 2020).

1.4 Defining Healthcare 5.0

The next stage of healthcare system development is Healthcare 5.0, which involves the use of new and advanced technologies into a fully interconnected and intelligent system that is intended to deliver predictive care, preventive care, and personalized care. In some way, this new paradigm brings Artificial Intelligence (AI), Internet of Medical Things (IoMT), robotics, digital twins, blockchain, and edge/cloud computing together. Unlike previous generations, which were more concerned with the amount of information that was gathered or automation, Healthcare 5.0 is about the ability to keep constant track of patients, plan treatment approaches in a more customized fashion, and predict diseases. With the help of continuous data collection from wearables and other IoMT sensors and cloud-based storage and immediate detailed analysis through AI, healthcare providers can now monitor patient health indicators in real time. Patient digital twins simulate the peculiarities of a patient in a virtual environment and allow medical professionals to test potential treatments and outcomes before they are actually implemented. Blockchain technology ensures that patients possess secure, unaltered records that can only be accessed by authorized officials, as data privacy and integrity are crucial concerns. Instead of patients being passive recipients of care, Healthcare 5.0 allows them to be active agents, offering them easy-to-use applications and dashboards and allowing them to follow their health and understand evidence-based plans of care (DeSalvo et al. 2017).

Table 1: Key Recent Research in Healthcare 5.0 (Haque and Rubya 2023).

Study	Focus Area	Key Contribution	Findings
Rajkomar et al. 2019	AI in Healthcare	Deep learning models for early disease detection	Improved early detection of sepsis and diabetic complications
Topol 2019	AI in Diagnostics	AI-based tools assisting in radiology and pathology	Enhanced diagnostic accuracy, especially in cancer and cardiovascular diseases
Alam et al. 2021	IoMT Applications	Survey of wearable devices and IoMT systems	Real-time remote monitoring of vitals, enabling early intervention
Islam et al. 2020	IoMT System Architecture	Edge computing in IoMT	Reduced latency, improved data processing efficiency
Yang et al. 2021	Robotics in Surgery	Use of surgical robots in minimally invasive surgeries	Reduced procedure times and complications
Xia et al. 2017	Blockchain for Medical Records	Blockchain framework for data security and privacy	Immutable, patient-controlled medical records
Tao et al. 2020	Digital Twins in Healthcare	Concept of patient digital twins for predictive simulation	Personalized treatment planning, preventive care enabled
Jiang et al. 2021	Ethical AI	Explainable AI for clinical decision-making	Addresses trust and ethical concerns in AI-based healthcare decisions

In spite of the advances, the current literature reveals that the main issues related to AI systems include data fragmentation, the absence of interoperability, ethical issues regarding the decision-making of AI, and expensive implementation. Some studies consider global standards and explainable AI models as the key to achieving trust and increased adoption in clinical practice (Jiang et al. 2021).

In brief, recent studies show a great drive toward the creation and adoption of Healthcare 5.0 technologies. Nevertheless, a lot of groundwork is still left to be done to combine these technologies into a cohesive, patient-centered ecosystem that is secure, interoperable, and sustainable. This chapter is inspired by recent research, and it presents a framework of how Healthcare 5.0 and real-life case studies should be understood, and how research should further be conducted in the future.

3. Background of Healthcare Evolution

Technology and evolving needs of society have radically transformed the concept of healthcare. Beginning with entirely manual functionality and ascending to highly connected smart technologies, each next stage of healthcare evolution was expected to enhance the efficiency of healthcare delivery, its accuracy, and accessibility. But every stage brought its own challenges and was likely to leave new loopholes as it closed the previous ones. This historical evolution can only be understood with the aim of appreciating the disruptive power of Healthcare 5.0, which attempts to establish a human-centric, predictive, and integrated care ecosystem.

This development is discussed in the following subsections that examine the main milestones in the development of care, as well as the shortcomings of old and current healthcare models.

3.1 Healthcare 1.0: The Analog Era

Healthcare 1.0 is the initial phase of healthcare, during which every operation was paper-based and manual. At this point, doctors used their own knowledge, direct interaction with the patient, and physical records as the sole basis of diagnosis and treatment of conditions they encountered. All patient histories were written out in heavy files, lab tests were done manually with no computer involvement, and medical decisions were made based on individual experience and medical education. The process of communication among healthcare providers was slow and could easily be distorted because of manual processing of data. The system was unsystematic and did not have any standardized medium or instrument for sharing data among hospitals or clinics. The treatment was purely reactive, and there were no preventative care mechanisms or early disease diagnosis mechanisms (Lu et al. 2022).

3.2 Healthcare 2.0: The Digital Beginnings

Healthcare 2.0 was the first period during which digital technologies began to be integrated into healthcare. With the launch of simple digital records and automation systems, hospitals were able to handle patient information more effectively than before the analog age. The administration started to digitalize (computerize) activities like patient appointments, billing, and storage of patient demographic information. In spite of the fact that this development enhanced the availability and systematization of information, these early digital solutions tended to be confined within separate hospital departments. This was caused by the absence of standardized formats and communication protocols, which resulted in disjointed data silos, thus making patient-sharing across institutions difficult. Clinical decision-making is one of the aspects where the use of technology did not help much because the judgment of physicians was taken as the primary instrument of diagnostics and predictive analysis (Maddikunta et al. 2022).

3.3 Healthcare 3.0: Integration of Electronic Health Records

Healthcare 3.0 ushered in a major shift toward much more systematic digitalization with the popularization of Electronic Medical Records (EMRs) and Electronic Health Records (EHRs). These systems contained more exhaustive electronic descriptions of patient medical histories consisting of laboratory test reports, medication orders, imaging tests, and clinical notes. EMRs also facilitated access to patient information for healthcare professionals more easily, which increased the speed of clinical workflow and reduced inefficiency in administration. There were some initial efforts to allow sharing of data across departments and hospitals, but interoperability was low. Healthcare 3.0 helped advance the accuracy of data collection, reduce duplications, and accelerate the documentation process. Yet, there were no system-wide predictive models or preventive measures, and clinical decisions remained largely reactive, i.e., based on symptoms observed at the point of care (Akundi et al. 2022).

3.4 Healthcare 4.0: The Era of Smart Connected Systems

Healthcare 4.0 is the next stage preceding Healthcare 5.0 with the introduction of smart devices, wearable health monitors, telemedicine platforms, and AI-assisted diagnostic support systems. IoMT devices enabled constant monitoring of the patient remotely, offering real-time statistics on vital parameters such as heart rate, blood glucose, blood pressure, and physical activity levels. Telemedicine sites widened the availability of medical care, particularly in rural or underserved areas, as it became possible to have a virtual appointment with an expert anywhere in the world. AI algorithms started to help clinicians with diagnostics by detecting patterns in medical images and recommending possible diagnoses and/or treatment plans based on previous data. Cloud computing provided centralized storage and real-time information accessibility across different devices and locations. However, despite these technological breakthroughs, Healthcare 4.0 was still heavily concerned with the implementation of technology rather than holistic care. Its care model was still responsive to disease symptoms instead of anticipating and averting health risks. The issue of data fragmentation and interoperability remained significant hurdles to the development of a truly connected system (Hamid 2024).

3.5 Limitations of Healthcare 4.0

There were numerous developments associated with Healthcare 4.0, but it also demonstrated a significant set of limitations that had to be addressed in the following evolutionary phase. Data fragmentation was one of the biggest problems. With various vendors converting medical devices and hospital systems into their own platforms, patient data tended to be siloed on different platforms, and it was hard to create a unified patient health record. Such fragmentation restricted the possibility of large-scale, long-term studies of the health course of a patient. Also, these systems were not interoperable. The majority of healthcare software tools were proprietary and closed systems that could not allow easy data exchange without advanced interfaces or human intervention. Also, although AI had been introduced in terms of diagnostics, the general care model was still reactionary. Respondents would act when their patients had symptoms instead of predicting them and missing the chance to intervene early. Lastly, Healthcare 4.0 was more inclined toward the efficiency of technology and data management rather than the human nature of healthcare. Even when devices gathered a lot of health information, patients were frequently given very little useful information that they could easily comprehend and act on. Healthcare was physician-focused, whereby technology became a complementary resource to care and not a partner (Ghobakhloo et al. 2022).

3.6 The Need for a Paradigm Shift

With upcoming changes in global healthcare, including the aging population, chronic diseases on the rise, disparities in healthcare access, and the recent lessons provided by the COVID-19 pandemic, a new paradigm must be adopted. From Table 2, it is evident that Healthcare 5.0 is expected to overcome the limitations of earlier models through predictive, preventive, and personalized healthcare services. It considers the convergence of advanced technologies as part of an autonomous patient-centered care system, decision support in real time, and proactive management of diseases. Healthcare 5.0 allows patients to gain control over their personal health, provides clinicians with effective predictive content models, and simultaneously facilitates the smooth flow of data while protecting data privacy and security as depicted in Figure 1. Also, It is not only the approach that will improve outcomes among patients but will also result in reduced costs, sustainability, and democratization of access to healthcare across different groups and regions (Kumar et al. 2023).

Table 2: Evolution of Healthcare Systems (Healthcare 1.0 → Healthcare 4.0) (Yüksel 2024).

Healthcare Phase	**Key Features**	**Technologies**	**Limitations**
Healthcare 1.0	Manual records, paper-based	None	Inefficient, prone to errors
Healthcare 2.0	Digital records, basic automation	EMRs	Fragmented systems
Healthcare 3.0	Integrated EMRs, early data analysis	Electronic Health Records (EHR)	Lack of predictive power
Healthcare 4.0	IoT devices, Telemedicine, AI support	Wearables, Telemedicine platforms, Cloud Computing	Data silos, poor interoperability
Healthcare 5.0	Human-centric, predictive, personalized care	AI, Digital Twin, Blockchain, Robotics, Edge Computing	Implementation cost, regulatory hurdles

4. Vision of Industry 5.0 and Its Relation to Healthcare 5.0

The operations of industries and service sectors in the world are changing significantly because of the increasing complexity of human needs and the rapid evolution of digital technologies. Industry 5.0 is an entirely new paradigm based on the interaction of intelligent machines and humans in order to create a more personalized, resilient, and sustainable industrial ecosystem. As healthcare is one of the most sensitive and vital sectors, it is likely to gain much from such a transformation. Healthcare 5.0 is the direct consequence of the overall concepts of Industry 5.0, with a straightforward implementation into the sphere of healthcare as highly personalized and predictive care with ethical considerations of human well-being and moral responsibility in the first place. The subsequent subsections discuss the essence of Industry 5.0 and its translation into a radical solution for healthcare provision (Ghobakhloo et al. 2022, Ivanov 2023).

systems that reduce the physical data center burden, and waste minimization in hospitals. Resilience is especially relevant during pandemics or natural disasters when healthcare systems must function even under the highest loads and significantly differing circumstances (Branke et al. 2016).

4.4 Linking Industry 5.0 to Healthcare 5.0

Healthcare 5.0 uses the concepts of Industry 5.0 to create a profound partnership between humans and sophisticated machines in the healthcare system. Nurses, physicians, and patients interact closely with smart technologies to enhance diagnosis, treatment, and prevention of diseases. The so-called digital twin, where a virtual image of the patient is created on the basis of real-time information about the patient delivered by IoMT devices, medical records, and predictive algorithms, is one of the technological facilitators of such a partnership. This digital twin simulates the present level of health of the patient and predicts the further evolution of the disease, taking into account the possibility of alternative treatment methods. It allows doctors to test interventions virtually and implement them with reduced risks and increased treatment individualization. Instead of treating patients according to standardized clinical guidelines, Healthcare 5.0 builds personalized, preventive, predictive, and participatory care plans based on real-time data-driven insights, which is why healthcare is more effective and patient-centered (Fang 2016).

4.5 Sustainable and Cost-Efficient Healthcare Solutions

One of the most relevant problems of healthcare today is the combination of the increasing cost of care and the increasing demand from the aging population and chronic diseases. Industry 5.0's sustainability vision can be transferred to healthcare solutions, which will lessen waste and expenditure, yet simultaneously enhance outcomes. A case in point is predictive maintenance of healthcare equipment, which reduces the risk of failure caused by malfunctioning medical equipment, and hence the cost of repair and downtime will be minimized. IoMT devices are energy-saving and extend battery life while reducing electronic waste. Cloud computing eradicates the need for physical hardware and thus reduces infrastructural costs incurred by hospitals and clinics. Besides, predictive analytics may be employed to sustain preventive care that may avoid expensive treatment through early detection. The strategy of sustainable design, smart resource use, and affordable technologies can help Healthcare 5.0 make high-quality care available and equitable even in low-resource settings (Redclift 2006).

5. Core Technologies Enabling Healthcare 5.0

The idea of Healthcare 5.0 symbolizes the overlapping of various new technologies which, when combined to work together, result in a highly intelligent, predictive, and patient-centered healthcare system. All these technological basics play a central role in transforming how healthcare is delivered, with the curve shifting toward proactive, individualized care rather than reactive care. The most important technologies discussed in the current section include Artificial Intelligence, IoMT, Robotics, Cloud and Edge computing, Digital Twins, and Blockchain, which enable the functionality of Healthcare 5.0 in the contemporary world to be efficient and sustainable (Madsen and Berg 2021).

5.1 Artificial Intelligence and Machine Learning

The essence of Healthcare 5.0 is centered on Artificial Intelligence (AI) and Machine Learning (ML), as the data processing system represents the brain of the system, enabling evidence-based decisions. Predictive analytics, powered by ML, can process substantial amounts of information acquired through various devices: wearable devices, EHRs, and laboratory results. These algorithms speedily and effortlessly detect concealed patterns and correlations that human beings may not observe and avert illnesses in their early stages. As an illustration, predictive models can forecast the risk of developing diabetes, cardiovascular diseases, or even cancer based on individual health parameters, enabling timely medical intervention. In addition to prediction, diagnostic algorithms can be helpful to radiologists and physicians as they automatically process medical images (CT scans, X-rays, MRI) and achieve the same or better accuracy than that of a human specialist. AI is also implemented in drug discovery, as it is precise in discovering possible compounds and predicting their activity based on molecular information, saving significant time and money in research. Another field where AI models are used is remote patient monitoring, where patients are tracked in real time and alerts are generated when their vital signs do not match regular patterns. The shift from one-size-fits-all care to incredibly customized lines of treatment is enabled by the use of AI and ML (Natarajan et al. 2023).

5.2 Internet of Medical Things (IoMT)

IoMT is imperative to the delivery of scalability in real-time data collection and seamless interrelation between patients, medical workers, and analytical systems. A wide variety of wearable devices are included in the IoMT, such as smartwatches, continuous glucose devices with real time monitoring, blood pressure cuffs, pulse oximeters, and smart inhalers that measure a variety of health-related variables in real time. These devices are programmed to continuously track patient data and securely provide it to cloud-based services, where it is analyzed with AI models. The advantages of IoMT include the ability to easily monitor patients at home so they do not have to come to the hospital, while simultaneously providing physicians with real-time data about their health. One can orchestrate home monitoring of heart failure patients using wearables such as ECGs to notify patients of arrhythmia or other heart rate abnormalities. Hospital environments also have connected medical devices that are part of IoMT, such as smart infusion pumps and smart beds, which make data-guided adjustments to treatment without human involvement. One of the significant improvements of the technology is the capability to provide proactive and real-time health services, making patients safer and less likely to be rehospitalized (Bhattacharya et al. 2021).

5.3 Robotics and Automation

Healthcare 5.0 is not limited to robotics and automation being applied in the manufacturing of medical equipment but also in the delivery of healthcare services. Surgical robots are one of the most popular uses, including the da Vinci Surgical System, which enables surgeons to perform extremely precise and minimally invasive surgeries. These robots minimize human error, enhance surgical outcomes, and reduce patient recovery time. Moreover, automated drug dispensers simplify the work of pharmacies, guaranteeing the correct distribution of medications, reducing human errors, and automating stock management. Routine work, such as transporting samples on hospital grounds, disinfecting rooms with UV light, or even assisting patients in rehabilitation or caring for elderly individuals, is also performed by robots. The Industry 5.0 vision may allow remote surgeries, enabling highly skilled surgeons to control robot arms in other hospitals over high-speed, low-latency networks. Not only has this enhanced access to specialized care in remote locations, but it has also provided opportunities to share treatment planning and consult medical expertise across the globe (Bhavin et al. 2021).

5.4 Cloud Computing and Edge Computing

Cloud computing is a compulsory component of Healthcare 5.0 since it provides on-demand and scalable computing and storage to manage massive amounts of medical information. Electronic Health Records (EHRs), IoMT, and AI-based models are transmitted and processed through cloud environments; therefore, patient information is available to healthcare professionals irrespective of their location anywhere on the planet. By using cloud infrastructure, data can be centralized, providing opportunities for coordinated care and large-scale data analysis, which helps to predict and design preventive care strategies. However, when it comes to real-time applications such as remote surgery or immediate wearable alerts, edge computing is necessary. Instead of transmitting all information to the cloud and waiting for a response, edge computing executes information on devices or nearby servers, removing significant latency and improving response time. For example, an on-body ECG device can process heart rhythm data locally and notify the patient or doctor of any critical situation without making cloud round-trip requests. Both cloud and edge computing offer an efficient and flexible data ecosystem, balancing large-scale data analysis and real-time processing to serve the objectives of Healthcare 5.0 (Tyagi et al. 2023).

5.5 Digital Twins in Healthcare

Digital Twin is an idea that is a spin-off of Industry 5.0, which is a virtual embodiment of a physical object that is updated in real time, contingent on constant inputs of information. Digital twins in healthcare are simulated versions of patients that replicate their current health condition and forecast the future health condition of a patient. Such models combine data provided by IoMT devices, genetic profiles, medical history, and lab results to create an accurate and dynamic simulation of the body and its response to treatments. Digital twins allow healthcare providers to perform simulations of what-if scenarios, enabling them to anticipate how a particular patient will react to a specific drug or treatment before conducting the experiment. As an example, a digital twin of a diabetic patient will be able to mimic how different dosages of insulin or lifestyle modifications will affect blood glucose levels, which will allow highly personalized care. Digital twins can transform healthcare into active prevention and treatment rather than a response to illnesses by predicting further progress and proposing timely measures. The fact that the digital twin is constantly synchronized with real-world data makes it an effective decision-support, medical training, and long-term health management tool (Yüksel 2024).

5.6 Blockchain for Data Security and Privacy

The safety of information and confidentiality of patients is one of the biggest concerns of modern healthcare. Healthcare 5.0 is a solution to these challenges through the application of blockchain technology, which provides a decentralized, immutable registry on which medical records can be stored. Every activity involving a patient health record is recorded in a blockchain transaction, and it is invariable and fully auditable. This ensures that medical data cannot be modified without requisite consent, preventing data violations and enhancing patient trust. One can also safely share information among hospitals, laboratories, and research institutions with patient consent and privacy being maintained through blockchain and the use of smart contracts. For instance, a patient can selectively choose access to his/her health

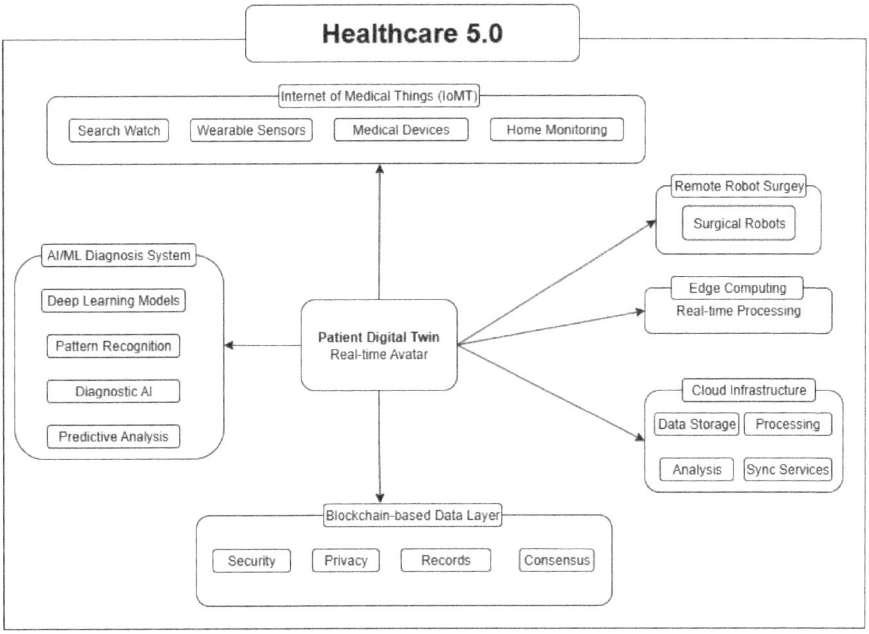

Figure 2: Healthcare 5.0 Core Technologies Framework (Naqvi et al. 2020).

ensuring that devices operate at their optimum and unnecessary replacements are minimized, reducing costs and e-waste. Edge computing eliminates the necessity of sending large-scale data to central servers, minimizing energy use and reducing network load. Cloud-based systems can also remove the need for expensive local hardware, as they allow resource sharing across institutions and provide scalable storage and processing power when required. In addition, Healthcare 5.0 reduces the total cost burden on healthcare systems by avoiding hospital readmissions through early disease detection and remote monitoring of patients. These solutions not only enhance the financial sustainability of healthcare providers but also promote environmental sustainability through reduced resource wastage (Hamid 2024).

6.6 Global Health Impact

In addition to personal patient care, Healthcare 5.0 has significant global health implications, especially in enhancing pandemic preparedness and enabling real-time health tracking on national and global scales. The significance of connected health systems was evident during the COVID-19 pandemic, when remote monitoring and data-sharing platforms allowed real-time tracking of infection spread and resource allocation. In the future, Healthcare 5.0 is expected to develop holistic networks of medical devices, patient records, and AI-based analytics that can detect outbreaks quickly, track contacts, and model disease transmission before it occurs. Anonymized data from IoMT devices and hospital records can help governments and other health agencies track community health trends, detect emerging health risks, and strategize responses. The ability to monitor situations in real time on a global scale allows the organization of cross-border actions and quicker reactions to pandemics and other major health crises. Moreover, with the democratization of access to advanced care technologies, Healthcare 5.0 contributes to reducing inequalities between developed and developing regions and makes high-quality healthcare more accessible worldwide (Jagadish et al. 2019). The comparison between Healthcare 4.0 and Healthcare 5.0 is discussed in Table 3.

Table 3: Comparison of Healthcare 4.0 vs Healthcare 5.0 (Pramod et al. 2022).

Feature	**Healthcare 4.0**	**Healthcare 5.0**
Focus	Technology-driven systems	Human-centric, personalized care
Data Handling	Mostly structured data	Real-time, multi-source data integration
Patient Involvement	Limited patient engagement	Active patient empowerment and participation
Predictive Capability	Basic predictive analytics	Advanced predictive and preventive care
Sustainability	Limited focus	Energy-efficient devices and waste reduction
Security	Fragmented security solutions	Blockchain-based data security and privacy
Interoperability	Limited standards	Global interoperability frameworks

7. Challenges and Barriers to Healthcare 5.0 Adoption

In spite of the huge potential of Healthcare 5.0 in transforming the way healthcare is delivered, there are several significant challenges and obstacles to implementation. These obstacles arise from technological, regulatory, ethical, financial, and cultural dimensions of health systems. Such barriers are vital to understand in order to develop realistic strategies to overcome them and ensure the best outcomes from Healthcare 5.0. The following subsections elaborate on the key challenges that must be addressed to enable effective implementation of this next-generation healthcare paradigm.

7.1 Data Privacy and Security

The security and confidentiality of data are among the top issues in the adoption of Healthcare 5.0 technologies. The amount of delicate health data collected by wearables, remote devices, cloud databases, and EHRs offers numerous points of vulnerability that can be exploited by malicious actors. The privacy of patients should be of the highest priority, since any form of data violation can have dire legal, monetary, and reputational consequences. Despite the fact that technological tools such as blockchain can ensure immutability and traceability, the safety of medical information must still be secured through effective encryption schemes, safe authentication procedures, and strong access control mechanisms. In most countries, several laws, including GDPR (General Data Protection Regulation) and HIPAA (Health Insurance Portability and Accountability Act), provide strict specifications regarding how patient information should be stored, transmitted, and shared. These rules pose complex technical and operational challenges when trying to implement systems that remain flexible and responsive in real time. Moreover, patients may feel uncertain about the use of their data in healthcare systems, and they will need the ability to understand and manage their personal information (Haque and Rubya 2023).

7.2 Regulatory Hurdles

The healthcare field is certainly one of the most controlled spheres due to the gravity of the provided services and the direct impact on human life. In different nations, healthcare technologies must remain safe, effective, and private under the strict compliance requirements of regulatory bodies. These laws often lag behind the accelerated pace of technological advancement, creating a gap between what can be technically done and what can be legally done. As an illustration, when AI-based diagnostic tools are introduced in hospitals, high-quality validation and clinical trials must be conducted to show that they are reliable and accurate. The technological innovations subject to regulatory oversight are not easily approved, making small innovators hesitant to enter the market. Likewise, IoMT devices are required to comply with high levels of safety and interoperability standards, which adds time and cost to development. This is further complicated by the absence of globally consistent regulations, with multinational healthcare providers operating under different rules in different regions. These barriers impede the implementation of Healthcare 5.0 technologies and introduce significant additional financial and administrative costs (Jagadish et al. 2019).

7.3 Interoperability Between Devices and Platforms

One of the background challenges of existing healthcare systems—and one of the greatest barriers to the implementation of Healthcare 5.0—is interoperability. The ecosystem consists of devices and platforms from different vendors that communicate through proprietary protocols, data formats, and interfaces. Therefore, integrating different medical devices (wearables, diagnostic equipment, imaging equipment) and software (EHR, analytics engines, patient portals) into a single, seamless system becomes highly complicated. Lack of interoperability leads to data silos whereby some parts of patient data cannot be easily moved and analyzed across institutions. This limits the possibilities of holistic patient treatment and large-scale predictive analytics. Initiatives to establish standardized ways of exchanging data (e.g., HL7 FHIR (Fast Healthcare Interoperability Resources)) are slow due to legacy systems and vendor unwillingness. Real interoperability can only be achieved through a combined effort by device manufacturers, software developers, healthcare providers, and regulatory bodies to develop common data and API standards and ensure secure communication protocols (Schiavoni et al. 2020).

7.4 Ethical Concerns in AI Decision-Making

The more Artificial Intelligence (AI) is involved in Healthcare 5.0, the more crucial ethical concerns arise, particularly regarding decision-making that directly affects patient care. AI models can often be viewed as black boxes, as they make decisions according to mathematical patterns that are intricate and difficult for humans to comprehend. Such non-transparency makes it challenging to justify and explain clinical decisions to patients and regulatory bodies. For example, if an AI model suggests a certain treatment plan or declares a patient as high-risk, patients and doctors may not be able to understand the reasoning behind it. This brings up the issue of accountability—who is responsible if the AI provides a wrong suggestion that causes harm? Additionally, bias in training data may lead to unequal treatment among various demographic groups, contributing to increased health disparities. Autonomous decision-making in healthcare environments requires ethical principles and transparent AI algorithms to ensure fairness, explainability, and trust in automated systems (Maddikunta et al. 2022).

7.5 High Costs of Implementing Advanced Technology

Implementing Healthcare 5.0 technologies requires significant investment in infrastructure, software development, hardware acquisition, and staff training.

Advanced technologies such as surgical robots, intelligent medical devices, Internet-of-Things wearables, and cloud services are expensive to purchase, maintain, and upgrade. Moreover, hospitals must allocate funding for IT infrastructure to support real-time data transmission, edge computing, and cloud storage. These costs are prohibitive for small and medium-sized healthcare providers, especially in developing nations. In addition to hardware and software expenses, employing and training personnel who can use advanced technologies, analyze large amounts of health data, and operate new systems creates an additional financial burden. These high costs make it unrealistic to implement Healthcare 5.0 rapidly and limit its adoption primarily to well-resourced hospitals and urban healthcare institutions, potentially widening the healthcare access gap (Ghobakhloo et al. 2022).

7.6 Resistance to Change in Traditional Healthcare Settings

Healthcare is a highly conservative sector in which workflow and hierarchical decision-making structures are deeply ingrained. These workflows are disrupted by the introduction of sophisticated technologies, creating uncertainty and resistance among healthcare professionals. Doctors, nurses, and administrators may be reluctant to use new tools due to fears that they will complicate their work, reduce professional autonomy, or threaten jobs. Moreover, older medical workers may resist adopting digital services, as they are accustomed to manual practices. Institutional inertia and the absence of appropriate change management strategies contribute to the slow implementation of Healthcare 5.0 practices. The transition must focus not only on technological readiness but also on helping healthcare professionals understand the benefits, receive training on using new devices, and participate in the design and implementation of these systems (Fang 2016). Key Challenges and their potential solutions are mentioned in Table 4.

Table 4: Summary of Key Challenges and Potential Solutions (Ivanov 2023).

Challenge	**Description**	**Potential Solution**
Data Privacy & Security	Risk of data breaches	Blockchain for immutability
Regulatory Hurdles	Different country regulations	Global interoperability standards
Interoperability	Fragmented platforms	HL7 FHIR implementation
Ethical AI	Black-box decision-making	Explainable AI (XAI)
High Costs	Expensive infrastructure	Energy-efficient IoMT designs
Resistance to Change	Lack of trust in new tech	Training & awareness programs

8. Case Studies and Real-World Applications

Healthcare 5.0 is not just a theoretical notion but is being implemented across the globe in the form of pilot programs and large-scale implementations by governments, healthcare providers, and private-sector companies. These practical examples demonstrate that advanced technologies such as Artificial Intelligence (AI), Internet of Medical Things (IoMT), blockchain, and digital twins are transforming healthcare

9. Future Outlook and Research Directions

With Healthcare 5.0 developing further, the future is full of opportunities for new technologies and emerging research directions. The future perspective is to incorporate next-generation communication, improved computing technologies, and smarter automation to further improve healthcare delivery. Simultaneously, there are a number of crucial research spheres that must be addressed to overcome current obstacles and provide continuous, responsible, and efficient healthcare innovation. The next subsections discuss some of the emerging trends and potential areas of research that will define the next stage of Healthcare 5.0 (Taj and Zaman 2022).

9.1 Emerging Trends: 6G and Future Communication Technologies

The development of communication technologies is one of the crucial factors in making the next generation of Healthcare 5.0 applications possible. As modern remote healthcare systems rely on 4G and 5G networks, the new 6G communication technology proposes ultra-high bandwidth, incredibly low latency, and the ability to connect enormous numbers of devices. These attributes will be essential for future health applications such as real-time remote surgeries, immediate delivery of ultra-high-resolution medical images, and seamless integration of hundreds of IoMT devices in hospitals and home care. To illustrate, during remote robotic surgeries, latency of more than a few milliseconds may hamper surgical accuracy and patient safety. 6G will allow healthcare professionals to carry out highly accurate remote surgeries and ensure immediate feedback. Moreover, 6G networks will enhance connectivity in underserved and rural areas where conventional broadband networks are limited, allowing equal access to high-quality healthcare services. This innovation in communications infrastructure will play a major role in supporting a fully integrated and predictive Healthcare 5.0 ecosystem (Taj and Zaman 2022, Tyagi et al. 2023).

9.2 Integration of Quantum Computing in Healthcare Simulations

The other frontier is quantum computing, which has colossal potential to transform Healthcare 5.0. Compared to classical computers, quantum computers work with quantum bits (qubits), enabling computation of very intricate issues within a fraction of the time. Advanced drug discovery and biological simulations remain among the most promising areas of quantum computing in healthcare. Traditional supercomputers may require weeks or months to simulate molecular interactions and protein folding of such complexity. Quantum computers can drastically accelerate these simulations, enabling researchers to examine molecular interactions more accurately and far more efficiently. Moreover, quantum machine learning algorithms can be used to create more accurate predictive models of disease progression and patient risk assessment, particularly in highly complex, high-dimensional medical datasets. Although quantum computing is not yet widely used in medical practice, it is expected to become a major breakthrough in drug creation, personalized medicine, and predictive diagnostics in the coming decades (Ávila-Gutiérrez et al. 2022).

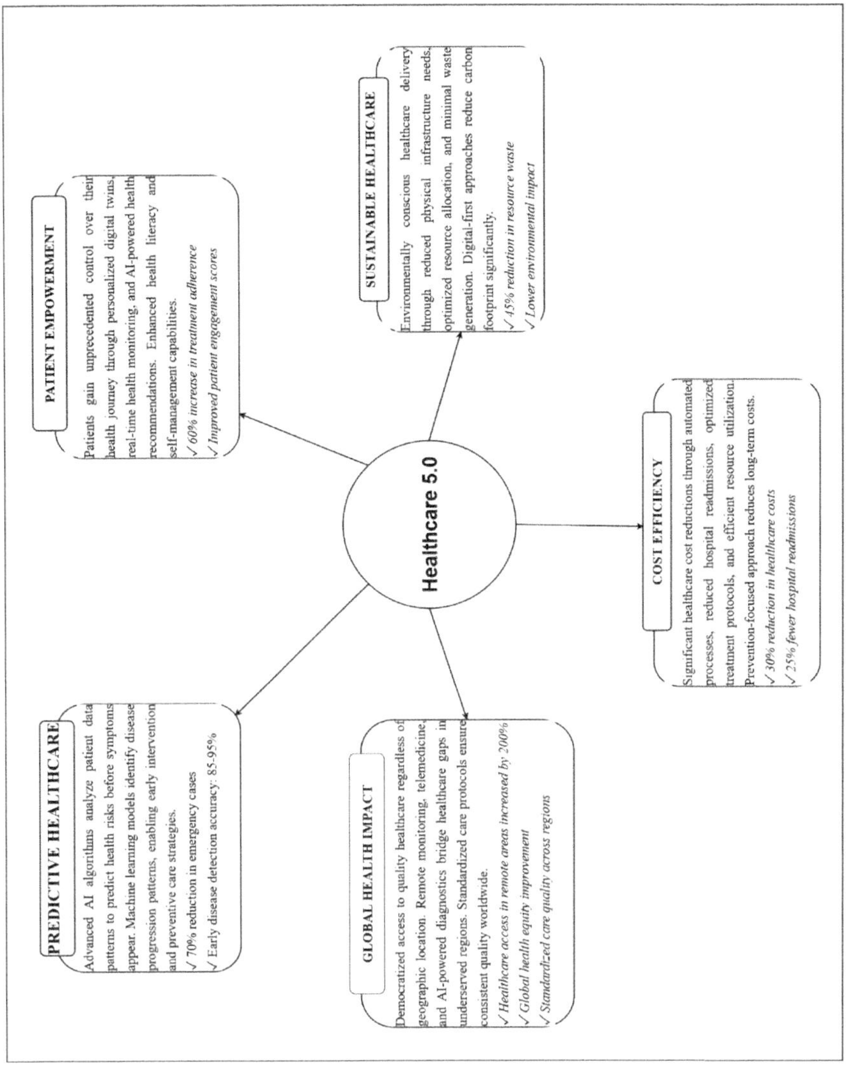

Figure 3: Future Healthcare 5.0 Technology Roadmap (Tyagi et al. 2023).

9.3 Advanced Robotics for Elderly Care

The growing needs of the aging population are creating an increased demand for innovative solutions to support independent living and enhance the quality of life of older adults. One promising sphere within Healthcare 5.0 is advanced robotics, expected to meet this demand by offering assistive and interactive robotic systems. In contrast to industrial robots designed for automation, healthcare robots are expected to interact empathetically with humans by reminding patients to take medicine, assisting with mobility, supporting daily activities, and providing companionship. Notably, projects such as RIBA-II (Robot for Interactive Body Assistance), developed in Japan, can lift patients and help with physical therapy exercises under medical supervision. Future studies aim to make these robots more flexible, equipped with natural language processing and emotional intelligence, enabling seamless assistance in elderly care homes. These robots can continuously monitor health indicators and alert medical personnel in emergencies, thereby promoting self-sufficiency, safety, and rapid response when needed (Yücebalkan 2020).

9.4 Research Areas: Explainable AI in Healthcare Decisions

The transparency of algorithms that facilitate decision-making is one of the most important concerns of AI implementation in healthcare, often referred to as the black box problem. This creates issues of trust, ethics, and regulatory compliance, especially when AI recommendations influence diagnosis and treatment. Explainable AI (XAI) has therefore become a significant area of interest, focusing on simplifying AI models so they are comprehensible to physicians and patients. XAI seeks to develop algorithms that can provide understandable explanations for their predictions or recommendations. For example, an XAI system would not only raise an alarm about a potential high risk of cardiac disease but also identify specific contributing factors such as abnormal ECG readings, elevated cholesterol, or family history. This builds trust, enhances clinical decision-making, and supports regulatory compliance. Future studies will focus on developing standardized frameworks and metrics for explainability, ensuring interpretability is not achieved at the cost of predictive accuracy (Nair et al. 2021).

9.5 Research Areas: Cross-Border Interoperability Standards

As healthcare becomes increasingly globalized, and medical devices, applications, and patient data move across borders, the lack of interoperability standards becomes a major barrier. To ensure the smooth flow of healthcare data between countries and systems, research is needed to establish international frameworks and open data standards enabling seamless data exchange. This is especially critical for pandemic preparedness, where real-time information sharing between nations is essential. On-going efforts such as the HL7 FHIR (Fast Healthcare Interoperability Resources) standard have made progress but still face obstacles in adoption, including lack of integration with legacy systems and conflicts with jurisdiction-specific privacy legislation. Future research will focus on creating adaptive and modular interoperability solutions that allow co-existence of old and new systems while complying with data sovereignty and privacy laws. This aims to establish a global health information ecosystem in which authorized providers can access complete and up-to-date patient information whenever needed to ensure continuity of care and informed decision-making in any circumstance (Zizic et al. 2022).

9.6 Research Areas: Energy-Efficient IoMT Devices

Energy efficiency is a new challenge in the proliferation of IoMT devices in Healthcare 5.0. Continuous supervision requires devices to remain powered on and to transmit real-time data, which increases energy usage and often requires batteries to be recharged or replaced more frequently. This cannot be sustained economically and environmentally, especially the implanted medical devices and remote sensors in the rural areas where power infrastructures are unsatisfactory. Recent research has focused on developing ultra-low-power sensors, energy harvesting (e.g., converting body heat or ambient light into power), and edge computing (to perform calculations on the sensor) to shrink the energy burden and reduce transmission energy costs. Data transmission protocols can also be optimized to transmit only necessary or compressed data to minimize energy consumption without reducing monitoring capabilities. Energy-efficient IoMT designs are essential to enable long-term, cost-effective, and environmentally friendly healthcare monitoring systems and support the long-term sustainable vision of Healthcare 5.0 (Zizic et al. 2022).

10. Conclusion

Healthcare 5.0 is not a paradigm exclusively concerned with the introduction of advanced technologies. Its future vision is to create healthcare systems that are not only highly efficient and data-driven but fundamentally human-centric, personalized, and sustainable. The chapter has addressed the evolution of healthcare from the paper-based systems of Healthcare 1.0 to the smart, connected, and predictive systems of Healthcare 5.0 that operate with the power of Artificial Intelligence (AI), the Internet of Medical Things (IoMT), blockchain, cloud computing, digital twins, and robotics. The overlap of these technologies creates a harmonious healthcare ecosystem that continuously adapts to the needs of each patient and promotes clinical decision-making, improved patient interaction, and operational sustainability.

In the discussion, we have observed how the field of customized medicine, propelled by predictive analytics and genetic profiling, enables personalized treatments to reach new heights, enhancing patient outcomes and minimizing unnecessary interventions. Telehealth and remote medical solutions have the potential to democratize care delivery by increasing access to medical care in rural and underserved locations. The paradigm of reactive treatment can be transformed through predictive and preventive healthcare models that support early-stage intervention enabled by real-time analytics and digital twin tools. Furthermore, the addition of blockchain ensures the safety, transparency, and immutability of patient data, fostering trust in digital healthcare ecosystems. Using energy-saving equipment

Schiavoni, R., Monti, G., Piuzzi, E., Tarricone, L., Tedesco, A., De Benedetto, E. et al. (2020). Feasibility of a wearable reflectometric system for sensing skin hydration. *Sensors*, 20(10): 2833. https://doi.org/10.3390/s20102833

Schuh, G., Potente, T., Wesch-Potente, C., Weber, A. R., and Prote, J.-P. (2014). Collaboration mechanisms to increase productivity in the context of industrie 4.0. *Procedia CIRP*, 19: 51–56. https://doi.org/10.1016/j.procir.2014.05.016

Stavropoulos, P., and Mourtzis, D. (2022). Digital twins in industry 4.0. In *Design and Operation of Production Networks for Mass Personalization in the Era of Cloud Technology* (pp. 277–316). Elsevier. https://doi.org/10.1016/B978-0-12-823657-4.00010-5

Taj, I., and Zaman, N. (2022). Towards Industrial Revolution 5.0 and explainable artificial intelligence: challenges and opportunities. *International Journal of Computing and Digital Systems*, 12(1): 285–310. https://doi.org/10.12785/ijcds/120124

Thieme, A., Belgrave, D., and Doherty, G. (2020). Machine learning in mental health. *ACM Transactions on Computer-Human Interaction*, 27(5): 1–53. https://doi.org/10.1145/3398069

Tiwari, S., Bahuguna, P. C., and Walker, J. (2022). *Industry 5.0* (pp. 59–73). https://doi.org/10.4018/978-1-7998-8497-2.ch004

Tyagi, A. K., Dananjayan, S., Agarwal, D., and Thariq Ahmed, H. F. (2023). Blockchain—Internet of Things Applications: Opportunities and Challenges for Industry 4.0 and Society 5.0. *Sensors*, 23(2): 947. https://doi.org/10.3390/s23020947

Wang, X., Markert, C., and Sasangohar, F. (2023). Investigating popular mental health mobile application downloads and activity during the COVID-19 pandemic. *Human Factors: The Journal of the Human Factors and Ergonomics Society*, 65(1): 50–61. https://doi.org/10.1177/0018720821998110

Xu, X., Lu, Y., Vogel-Heuser, B., and Wang, L. (2021). Industry 4.0 and Industry 5.0—Inception, conception and perception. *Journal of Manufacturing Systems*, 61: 530–535. https://doi.org/10.1016/j.jmsy.2021.10.006

Yücebalkan, B. (2020). Endüstri 4.0’Dan Endüstri 5.0’A Geçiş Sürecine Genel Bakiş. *I Association*, 5(9): 241–250. https://doi.org/10.46872/pj.181

Yüksel, O. (2024). An overview of healthcare services in the Industry 5.0 era. *International Journal of Engineering and Innovative Research*, 6(3): 222–237. https://doi.org/10.47933/ijeir.1537942

Zaffiri, L., Gardner, J., and Toledo-Pereyra, L. H. (2012). History of Antibiotics. From Salvarsan to Cephalosporins. *Journal of Investigative Surgery*, 25(2): 67–77. https://doi.org/10.3109/08941939.2012.664099

Zizic, M. C., Mladineo, M., Gjeldum, N., and Celent, L. (2022). From Industry 4.0 towards Industry 5.0: A review and analysis of paradigm shift for the people, organization and technology. *Energies*, 15(14): 5221. https://doi.org/10.3390/en15145221

Index

About the Editors

Dr. Vivek Kumar Singh is an Assistant Professor at Sharda University with 13+ years of teaching, industrial, and research experience in Computer Science and Engineering. His expertise includes computer vision, machine learning, artificial intelligence, and image processing. He has published several SCI/Scopus-indexed journal papers, conference articles, book chapters, and CRC Press edited books in emerging technologies and smart healthcare. Dr. Singh also holds multiple published patents in intelligent systems, healthcare, and automation, reflecting his strong research contributions toward innovative AI-driven and sensor-enabled technological solutions.

Dr. Avinash Kumar Sharma is an Associate Professor at Sharda University with 19+ years of academic and research experience in Computer Science and Engineering. His expertise includes artificial intelligence, cloud computing, IoMT, machine learning, cybersecurity, and smart healthcare systems. He has authored numerous Scopus-indexed journal papers, conference papers, book chapters, and edited books with reputed publishers including IGI Global, Wiley, Bentham Science, and CRC Press. Dr. Sharma also holds multiple Indian patents and design patents, reflecting his strong contributions to innovation, intelligent systems, and emerging healthcare technologies.

Dr. Tuan Anh Nguyen is Senior Principal Research Scientist at the Vietnam Academy of Science and Technology, Vietnam. He received B.S. in Physics from Hanoi University in 1992, and Ph.D. in Chemistry from the Paris Diderot University (France) in 2003. He was a Visiting Scientist at Seoul National University (South Korea, 2004) and University of Wollongong (Australia, 2005). He then worked as Postdoctoral Research Associate and Research Scientist in the Montana State University (USA), 2006–2009. In 2012, he was appointed as the Head of the Microanalysis Department at Institute for Tropical Technology. His research activities include smart sensors, smart networks, smart hospitals, smart cities and advanced nanomaterials. He has edited more than 75 Elsevier, 13 CRC Press, 1 Springer, 1 RSC and 2 IGI Global books. He is the Editor-In-Chief of "Kenkyu Journal of Nanotechnology & Nanoscience".